I0063207

Crystal Structure of Electroceramics

Special Issue Editor

Stevin Snellius Pramana

MDPI • Basel • Beijing • Wuhan • Barcelona • Belgrade

MDPI

Special Issue Editor
Stevin Snellius Pramana
Newcastle University
UK

Editorial Office
MDPI AG
St. Alban-Anlage 66
Basel, Switzerland

This edition is a reprint of the Special Issue published online in the open access journal *Crystals* (ISSN 2073-4352) in 2017 (available at: http://www.mdpi.com/journal/crystals/special_issues/crystal_structure_electroceramics).

For citation purposes, cite each article independently as indicated on the article page online and as indicated below:

Author 1; Author 2. Article title. *Journal Name* **Year**, *Article number*, page range.

First Edition 2017

ISBN 978-3-03842-593-9 (Pbk)
ISBN 978-3-03842-594-6 (PDF)

Articles in this volume are Open Access and distributed under the Creative Commons Attribution license (CC BY), which allows users to download, copy and build upon published articles even for commercial purposes, as long as the author and publisher are properly credited, which ensures maximum dissemination and a wider impact of our publications. The book taken as a whole is © 2017 MDPI, Basel, Switzerland, distributed under the terms and conditions of the Creative Commons license CC BY-NC-ND (http://creativecommons.org/licenses/by-nc-nd/4.0/).

Table of Contents

About the Special Issue Editor

Stevin Snellius Pramana is a Lecturer in Chemical Engineering in the School of Engineering at Newcastle University, UK. He obtained his PhD from Nanyang Technological University, Singapore and continued to pursue his postdoctoral research career at Imperial College London, UK. He is primarily interested in understanding the crystallography and chemistry in the bulk and on the surface of energy conversion and storage materials in the form of defective, modulated, disordered and ordered crystals, by using a combination of material characterisation techniques, including electron microscopy, synchrotron and neutron scattering and spectroscopy methods.

Preface to "Crystal Structure of Electroceramics"

Electrical, optical, ionic and magnetic properties of ceramics are primarily dictated by their crystal structure. They can be improved by introducing impurities, creating long range order/short range order/disorder, engineering defects and utilising specific crystal anisotropy and orientation. This Special Issue is aimed at manuscripts focusing on the recent development of electroceramics and its relation to the crystallography, including the characterisation aspect.

In particular, the topic of interest covers the review paper on principles, difficulties and progress of crystal structure determination and refinement from powder diffraction, and ferroelectricity in binary crystals. Other interesting piezoelectric, electrical conducting, dielectric and ferroelectric ceramics are reported with various crystal structural characterisation and microscopy techniques.

Stevin Snellius Pramana

Special Issue Editor

crystals

MDPI

Article

Site Identity and Importance in Cosubstituted Bixbyite In$_2$O$_3$

Karl Rickert [1], Jeremy Harris [1], Nazmi Sedefoglu [2], Hamide Kavak [3], Donald E. Ellis [4] and Kenneth R. Poeppelmeier [1,*]

[1] Department of Chemistry, Northwestern University, Evanston, IL 60208, USA;
    karlrickert2011@u.northwestern.edu (K.R.); jeremyharris2013@u.northwestern.edu (J.H.)
[2] Department of Physics, Osmaniye Korkut Ata University, Osmaniye 80000, Turkey;
    nazmisedefoglu@gmail.com
[3] Physics Department, Cukurova University, Adana 01330, Turkey; hkavak@cu.edu.tr
[4] Department of Physics and Astronomy, Northwestern University, Evanston, IL 60208, USA;
    don-ellis@northwestern.edu
*   Correspondence: krp@northwestern.edu; Tel.: +1-847-491-3505

Academic Editor: Stevin Snellius Pramana
Received: 14 December 2016; Accepted: 6 February 2017; Published: 9 February 2017

Abstract: The bixbyite structure of In$_2$O$_3$ has two nonequivalent, 6-coordinate cation sites and, when Sn is doped into In$_2$O$_3$, the Sn prefers the "*b*-site" and produces a highly conductive material. When divalent/tetravalent cation pairs are cosubstituted into In$_2$O$_3$, however, the conductivity increases to a lesser extent and the site occupancy is less understood. We examine the site occupancy in the Mg$_x$In$_{2-2x}$Sn$_x$O$_3$ and Zn$_x$In$_{2-2x}$Sn$_x$O$_3$ systems with high resolution X-ray and neutron diffraction and density functional theory computations, respectively. In these sample cases and those that are previously reported in the M$_x$In$_{2-2x}$Sn$_x$O$_3$ (M = Cu, Ni, or Zn) systems, the solubility limit is greater than 25%, ensuring that the *b*-site cannot be the exclusively preferred site as it is in Sn:In$_2$O$_3$. Prior to this saturation point, we report that the M^{2+} cation always has at least a partial occupancy on the *d*-site and the Sn^{4+} cation has at least a partial occupancy on the *b*-site. The energies of formation for these configurations are highly favored, and prefer that the divalent and tetravalent substitutes are adjacent in the crystal lattice, which suggests short range ordering. Diffuse reflectance and 4-point probe measurements of Mg$_x$In$_{2-x}$Sn$_x$O$_3$ demonstrate that it can maintain an optical band gap >2.8 eV while surpassing 1000 S/cm in conductivity. Understanding how multiple constituents occupy the two nonequivalent cation sites can provide information on how to optimize cosubstituted systems to increase Sn solubility while maintaining its dopant nature, achieving maximum conductivity.

Keywords: bixbyite; indium oxide; transparent conducting oxide

1. Introduction

Bixbyite indium oxide is the basis of industrially important n-type transparent conducting oxides, which are in turn fundamental components of flat panel displays, touch screens, and solar cells [1–3]. Specifically, tin-doped indium oxide (ITO) is a highly effective transparent conducting oxide that derives its desirable conductivity from the formation of Frank-Köstlin clusters [4]. The concentration of these clusters are directly influenced by the solubility of tin in the bixbyite lattice. As higher conductivities are being demanded of transparent conductors, increasing the solubility of tin in the bixbyite structure is of interest.

One proven method to increase the solubility of tin in In$_2$O$_3$ is to perform a cosubstitution of M^{2+} and Sn^{4+} into In$_2$O$_3$, where M = Mg, Ca, Ni, Cu, Zn, or Cd, to form M$_x$In$_{2-2x}$Sn$_x$O$_3$ which maintains the bixbyite structure of In$_2$O$_3$. In so doing, solubility limits as high as *x* = 0.5 can be achieved, which is approximately double the amount of tin in ITO [5,6]. This does not double the conductivity of ITO, however, as the cosubstitution is not true doping because the tetravalent tin is counterbalanced by a divalent cation when replacing two trivalent cations. A complete counterbalancing effect would substantially decrease the conductivity when compared to ITO, but in some cases an inherent off stoichiometry in the divalent and tetravalent cation is observed, which results in a dopant effect and high conductivity [7].

In ITO, the dopant favors one of the two nonequivalent cation sites present in the bixbyite structure. Both sites are 6-coordinate and can be described as a cube with two anion positions vacant along either a body diagonal (*b*-site) or a face diagonal (*d*-site). Computational and experimental studies agree that the tin in ITO favors the *b*-site [8–10]. The *b*-site comprises 25% of the cation sites in the bixbyite structure, shown in Figure 1, which would inherently limit the solubility of tin, but the solubility limit is actually lower, resulting in mixed indium and tin occupancy of the *b*-site. The solubility limits of cosubstituted bixbyite, however, can occupy up to 50% of the cation sites and therefore must alter the *d*-site either exclusively or in combination with the *b*-site. The impact of multiple constituents (i.e., only Sn vs. M/Sn pairs) on the electronic properties has already drawn investigative interest, particularly for M = Zn, but the importance of site occupancy has not yet been tied to either the multiple constituents or the electronic properties [11,12]. The constituent identities and their site preferences have been linked to crystal structure, however, as exhibited by the cosubstitution of Zn/Ge pairs. In this instance, the system forms a single defined phase instead of a solid solution and departs from the bixbyite structure in order to provide a 4-coordinate site for Ge [13].

Figure 1. The unit cell (black lines) of bixbyite In_2O_3 as viewed along the *b* axis, showing the locations of the *b*-sites (opaque, orange, 25% of cation sites) and the *d*-sites (transparent, blue, 75%).

A thorough understanding of how multiple constituents occupy the two nonequivalent cation sites in bixbyite can provide information on how to optimize the cosubstituted systems to increase tin solubility while maintaining its dopant nature. Herein, the cosubstituted bixbyite system $M_xIn_{2-2x}Sn_xO_3$, with M = Mg or Zn, is evaluated and the site occupancies and electronic properties are compared to each other and reported cosubstituted systems. The $Zn_xIn_{2-2x}Sn_xO_3$ system (commonly referred to as either ZITO or IZTO) is widely studied as a transparent conductor and the electronic properties have been previously reported, but here it is investigated with a computational approach that examines local site structure [6,11,14–19]. $Mg_xIn_{2-2x}Sn_xO_3$, in contrast, has been the subject of a single bulk study and a thin film study and here is the subject of a more thorough structural and property study [5,20]. Local coordination environments and site occupancies are difficult to confidently determine with experimental techniques, which average sites in a bulk sample, and thus computational approaches are more informative in these approaches. Both methods are presented here and suggest that the *b*-site is still the more important site for the tetravalent cation and thus the conductivity of these materials.

2. Results

2.1. Solid Solution Characterization

The lattice parameters for as-synthesized $Mg_xIn_{2-2x} - Sn_xO_3$ ($x \leq 0.3$) are determined by X-ray diffraction (XRD) and the lattice parameters for $Zn_xIn_{2-2x} - Sn_xO_3$ ($x \leq 0.5$) are calculated via density functional theory (DFT). Both sets of lattice parameters are provided in Figure S1 and the XRD patterns used to calculate the parameters for $Mg_xIn_{2-2x} - Sn_xO_3$ are shown in Figure S2. The lattice parameters exhibit linear decreases, following Vegard's Law, as a result of the smaller radii of Mg/Sn (0.72 Å/0.69 Å) and Zn/Sn (0.74 Å/0.69 Å) pairs compared to that of In/In (0.80 Å) pairs [21–23]. The $Mg_xIn_{2-2x} - Sn_xO_3$ lattice parameters closely match those that have been previously reported for the system, but the $Zn_xIn_{2-2x} - Sn_xO_3$ lattice parameters are systematically expanded, a somewhat common occurrence when DFT results are compared with experimental results [24,25]. The solubility limit of the Mg/Sn pairing (0.3) is slightly higher than previous reports (0.25) [5]. Surpassing the solubility limit in $Mg_xIn_{2-2x} - Sn_xO_3$ produces the easily identified SnO_2 and MgO phases. Off stoichiometry trials (i.e., where [Mg] $\neq$ [Sn]) were attempted, but were unsuccessful in creating the Sn rich phases reported via thin film synthesis [20].

2.2. $Mg_{0.1}In_{1.8}Sn_{0.1}O_3$ Structural Refinement

A joint Rietveld refinement of synchrotron XRD and time-of-flight neutron diffraction (TOF-ND) data for $Mg_{0.1}In_{1.8}Sn_{0.1}O_3$ ($x = 0.1$) is considered. The overall bixbyite structure is maintained and the goodness-of-fit parameters for a joint Rietveld refinement of each probable occupancy pattern are provided in Table 1. The Rietveld refinement and difference patterns are provided in Figure S3. As can be observed, all of the occupancy patterns display similar fits, with differences in total R_{wp} values being less than 0.0012. The χ^2 and Bragg factors also display similar values. One factor that negatively impacts the goodness of fit parameters for the XRD refinement is a systematic, asymmetric peak shape with tails at lower d values. Such a shape is typically associated with stacking or deformation faults. The fits discussed here do not account for this asymmetry, as the exact cause is unclear and thus an appropriate model cannot be applied. Even without accounting for this asymmetry, the fits are reasonable but maintain residual intensity on each reflection. Using Hamilton's R test with 45 refined parameters, 545 XRD reflections, and 1080 ND reflections, the difference between the R_{wp} values ≥ 0.2208 and the R_{wp} values ≤ 0.2199 is statistically significant at the 0.5% level. That is, the hypothesis that the fits without Mg on the d-site are correct is rejected [26]. The differences between the remaining models, however, are not statistically significant. The qualitative assessment of the difference patterns for these fits do not offer assistance in selecting a site. These data suggest that a "correct" structural model must contain Mg on the d-site, but the exact quantity is unknown and the Sn location is inconclusive. It is possible that this ambiguity arises from the aforementioned asymmetric peak shapes that are not accounted for in the refinement, but it is also possible that introducing a model to handle the asymmetry would decrease the ability to accurately discern occupancies, maintaining the inconclusive results.

Table 1. Goodness of fit parameters for joint X-ray diffraction (XRD) and ND Rietveld refinement of $Mg_{0.1}In_{1.8}Sn_{0.1}O_3$ with different cation occupancies. Indium occupies the remaining site percentages.

Mg Location (%)		Sn Location (%)		R_{wp}			χ^2	R_{F^2}	
b-site	d-site	b-site	d-site	Total	XRD	ND		XRD	ND
20	0	0	6.67	0.2209	0.2694	0.0423	24.32	0.0947	0.1442
0	6.67	20	0	0.2198	0.2680	0.0426	24.09	0.0917	0.1528
20	0	20	0	0.2208	0.2692	0.0433	24.30	0.0940	0.1420
0	6.67	0	6.67	0.2198	0.2679	0.0431	24.08	0.0913	0.1544
10	3.33	10	3.33	0.2199	0.2681	0.0420	24.09	0.0898	0.1482

2.3. $Zn_xIn_{2-2x}Sn_xO_3$ Formation Energy Computations

The formation energies of $Zn_xIn_{2-2x}Sn_xO_3$ for different values of x are calculated and provided in Figure 2a. As the amount of substitutes increases, the favorability of the reaction decreases and the results are prone to a higher variability. Even with this increased variability, the data set displays a good linear fit for $x > 0$ until the final data point ($x = 0.50$) which changes the R^2 value from 0.9817 to 0.9515. Note that previous reports place the experimentally determined solubility limit as $x \leq 0.4$ in $Zn_xIn_{2-2x}Sn_xO_3$ [6]. The higher variability is a result of the greater possible variety of site occupancies of the substitutes that are considered. These different occupancies are considered in Figure 2b for a single Zn/Sn pair. A shorter Zn-Sn separation difference is the most favored configuration, regardless as to the specific site occupancies. The relative favorability between the different site occupancies is maintained with distance and the overall most favorable occupancy pattern is Zn present on the *d*-site, Sn present on the *b*-site, and the Zn and Sn present in neighboring positions.

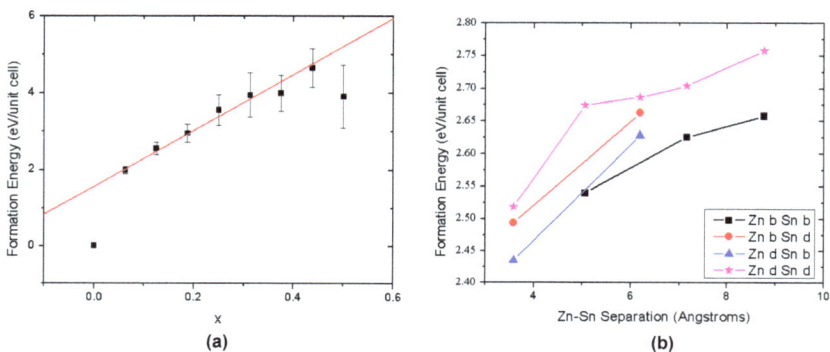

Figure 2. Formation energies as a function of (**a**) cosubstitution amounts of $Zn_xIn_{2-2x}Sn_xO_3$ and (**b**) Zn-Sn separation for different site occupancies (i.e., "Zn b Sn b" represents Zn and Sn present on the b site). The linear fit in (**a**) is for $0 < x < 0.5$. All values in (**a**,**b**) are normalized to the -449.832 eV of In_2O_3 ($x = 0$), thus the greater values have less favored formation energies.

2.4. $Mg_xIn_{2-2x}Sn_xO_3$ Conductivity and Band Gap Measurements

The original discovery of the $Mg_xIn_{2-2x}Sn_xO_3$ solid solution assessed the optical properties of $x = 0.05$ and 0.25 and the temperature-dependent conductivity of $x = 0.05$. From this preliminary characterization, the band gap is known to be similar to that of ITO, but the conductivity is approximately one order of magnitude lower than that of ITO [5]. The Seebeck coefficients of the as-synthesized materials (see Figure S4) have a negative parity, identifying the $Mg_xIn_{2-2x}Sn_xO_3$ system as n-type. In the original investigation, the samples were air quenched, but reduction treatments of n-type TCOs are common practices for increasing oxygen defects and therefore the conductivity and had not yet been performed. The room temperature conductivity of $Mg_xIn_{2-2x}Sn_xO_3$ before and after such a reduction procedure is shown in Figure 3 [27]. The conductivity of as-synthesized $Mg_xIn_{2-2x}Sn_xO_3$ was previously reported to be on the order of $10^0 \, \Omega \cdot cm$ (equivalent to 10^0 S/cm), but here, the conductivity of the as-synthesized $Mg_xIn_{2-2x}Sn_xO_3$ ranges from 5.8×10^1 S/cm^{-1} to 1.56×10^2 S/cm^{-1}, which are orders of magnitude greater. Furthermore, upon reduction, the conductivity of $Mg_xIn_{2-2x}Sn_xO_3$ can reach as high as 1064 S/cm, which makes that specific composition ($x = 0.05$) of potential interest as a commercial TCO material. Most notable is the severe drop in reduced conductivity from $x = 0.05$ to 0.1. Although conductivity typically decreases with decreasing In content in cosubstituted In_2O_3, a decrease of such magnitude is remarkable [11]. Powder XRD of the reduced samples do not exhibit any structural alterations, such as the emergence of secondary phases, that would account for the increased conductivity (relative to the pre-reduction conductivity).

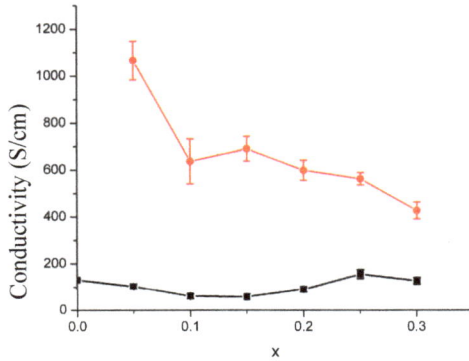

Figure 3. The conductivity of $Mg_xIn_{2-2x}Sn_xO_3$ before (squares) and after (circles) reduction. Error bars may be hidden by the data points. The point at $x = 0$ (pure In_2O_3) is reproduced from Reference [27], which follows a similar sample preparation.

With the favorable conductivity, the optical properties of the $Mg_xIn_{2-2x}Sn_xO_3$ solid solution are of interest and have also been assessed. Prior to reduction, pellets of $Mg_xIn_{2-2x}Sn_xO_3$ are pale yellow. After reduction, the same pellets are blue-gray. The optical band gaps of these materials are displayed in Figure 4 and, as can be observed, exhibit a ~0.1 eV increase upon reduction. Such an increase is common in n-type TCOs, as the reduction procedure increases the number of carriers, as evidenced by the increased conductivity, which enables a Burstein-Moss shift to occur, elevating the value of the optical band gap [28,29]. The fact that these materials are still n-type is confirmed by the Seebeck coefficients of the reduced samples (Figure S4), as they are negative. As mentioned in the discovery of the $Mg_xIn_{2-2x}Sn_xO_3$ system, the optical properties compare favorably with those of ITO. Thus the $Mg_xIn_{2-2x}Sn_xO_3$ solid solution can display both a high transparency and conductivity.

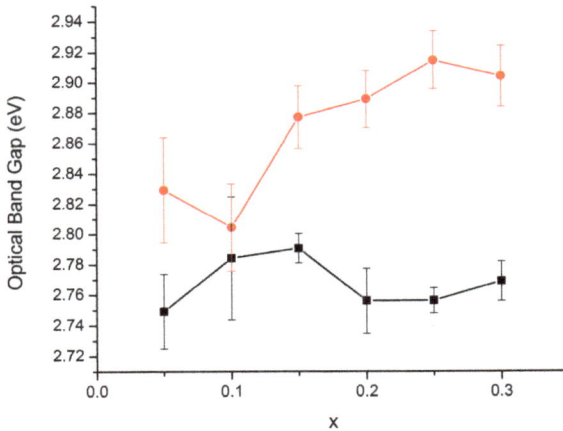

Figure 4. Optical band gaps of the $Mg_xIn_{2-2x}Sn_xO_3$ solid solution before (squares) and after (circles) reduction.

3. Discussion

The greatest question facing a complete structural solution of the $Mg_xIn_{2-2x}Sn_xO_3$ system is the location(s) of the substituting elements. As demonstrated in Table 1, the joint Rietveld refinement does not provide a definite occupancy of the structure. The structurally characterized members of the $M_xIn_{2-2x}Sn_xO_3$ (M = Ca, Ni, Cu, Zn, or Cd) family can be considered for guiding which of the aforementioned occupancy models is most accurate and the site occupancies of those that have been structurally characterized (M = Ni, Cu, Zn) are provided in Table 2. In all cases where Rietveld refinement of X-ray and neutron data were used, (1) a substantial amount of the M^{2+} substitute is present on the d-site; (2) there is some Sn^{4+} present on the b-site; and (3) the majority of M^{2+} is present on the site with the minority of Sn^{4+}. Unlike these overarching trends, relative site concentrations vary significantly, with a majority of M^{2+} being able to be present on either the b-site (M = Ni) or the d-site (M = Cu, Zn) and Sn^{4+} being entirely confined to the b-site (M = Cu) or nearly confined to the d-site (M = Ni). In the similar $In_{2-x}X_{2x/3}Sb_{x/3}O_3$ (X = Zn or Cu) system, however, Sb^{5+} only occupies the b-site, which is similar to the occupancy of Sn^{4+} in ITO and one of the best fits for $Mg_{0.1}In_{1.8}Sn_{0.1}O_3$ [9,10,30] With the similar goodness-of-fit parameters for the different occupancy models of $Mg_{0.1}In_{1.8}Sn_{0.1}O_3$, and the lack of a distinguishing trend in the $M_xIn_{2-2x}Sn_xO_3$ family, the most reliable conclusion from this structural investigation is that there must be some Mg on the d-site in $Mg_{0.1}In_{1.8}Sn_{0.1}O_3$.

Table 2. Site occupancies of reported cosubstituted bixbyite phases. Indium occupies the remaining site percentages.

Material	M Location (%)		Sn Location (%)		Method
	b-site	d-site	b-site	d-site	
$Cu_{0.275}In_{1.45}Sn_{0.275}O_3$	14	14	55	0	Rietveld Refinement—Neutron and X-ray [5]
$Ni_{0.5}InSn_{0.5}O_3$	65	11	2	33	Rietveld Refinement—Neutron and X-ray [5]
$Zn_{0.1}In_{1.8}Sn_{0.1}O_3$	No ordering, randomly distributed				Extended X-ray Absorption Fine Structure [14]
$Zn_{0.2}In_{1.6}Sn_{0.2}O_3$	No ordering, randomly distributed				Extended X-ray Absorption Fine Structure [14]
$Zn_{0.25}In_{1.5}Sn_{0.25}O_3$	18	11	50	0	Rietveld Refinement—Neutron and X-ray [5]
$Zn_{0.3}In_{1.4}Sn_{0.3}O_3$	No ordering, randomly distributed				Extended X-ray Absorption Fine Structure [14]
$Zn_{0.4}In_{1.2}Sn_{0.4}O_3$	No ordering, randomly distributed				Extended X-ray Absorption Fine Structure [14]

As shown in Table 2, the M = Zn system has been structurally characterized previously, but the reported occupancies are in conflict. Powder XRD studies suggest that Sn^{4+} only occupies the b-site, but additional studies with extended X-ray absorption fine structure (EXAFS) analysis have shown that both Zn and Sn are statistically distributed over the two available sites, with the probable formation of Sn clusters [5,14,15]. These contradicting results and the difficulty in definitively determining the site occupancies in $Mg_xIn_{2-2x}Sn_xO_3$, which is likely exacerbated by a microstructure disorder that results in the asymmetric peak shapes, suggest that an approach that avoids these difficulties is needed. Computational chemistry offers such a solution and the formation energies of different occupational models of $Zn_xIn_{2-2x}Sn_xO_3$ are calculated to seek a guiding principle and are provided in Figure 2b. Regardless as to the site occupancies, the higher the degree of separation, the lower the favorability of the formation, suggesting short range ordering may be present. The most favored points are when Zn and Sn are separated by 3.5769 Å, which translates to Zn and Sn being present in adjacent cation sites. Of the configurations that can have Zn and Sn adjacent, the mixed configurations, specifically when Zn is on the d-site and Sn is on the b-site, are most favored. These findings corroborate earlier reports of the stability of Zn on the d-site and Sn on the b-site. In the prior study, however, substitution on solely the b- or d-site was not mentioned [12]. This also fits with what is currently known experimentally, as Sn prefers the b-site in ITO and M^{2+} is always present on the d-site in $M_xIn_{2-2x}Sn_xO_3$. The inverse of these trends, when Sn is only present on the d-site and Zn is only present on the b-site are far less favored. Furthermore, the least favored configuration is when the b-site is not substituted, again suggesting that ITO's trends are still apparent in the cosubstituted systems. Interestingly, the stability switches to complete b-site substitution once the larger separations are considered. This suggests that local

M^{2+}/Sn^{4+} interactions stabilize the structure, but the b-site is energetically favored if the interactions are negligible.

4. Materials and Methods

4.1. $Mg_xIn_{2-2x}Sn_xO_3$ Sample Preparation

Samples of $Mg_xIn_{2-2x}Sn_xO_3$ (x = 0.05, 0.1, 0.15, 0.2, 0.25, and 0.3) were prepared by mixing stoichiometric quantities of MgO, In_2O_3, and SnO_2 via a Fritsch planetary ball mill using agate media at 600 rpm for 4 cycles of 15 min, with pauses of 5 min between cycles. MgO was subjected to thermogravimetric analysis prior to this preparation and no carbonates or hydroxides were detected. Each sample was pressed into six, 13 mm in diameter cylindrical pellet under 16000 psi, buried in sacrificial powder in separate nested alumina crucibles, and heated to 1300 °C at a rate of 5 °C per min. The temperature was held at 1300 °C for 28 h before being cooled to 25 °C at a rate of 5 °C per min. Reduced samples were prepared on beds of sacrificial powder in alumina boats and ramped to 500 °C at a rate of 5 °C per min, dwelled at 500 °C for 7 h, and cooled to room temperature at a rate of 5 °C per min under a 5% hydrogen atmosphere (argon balance).

4.2. $Mg_xIn_{2-2x}Sn_xO_3$ Sample Characterization

Structure studies were performed with X-ray and neutron diffraction. A sample of $Mg_{0.1}In_{1.8}Sn_{0.1}O_3$ was prepared as a 2.54 cm cylindrical pellet and treated to the same procedure as above, but was not subjected to the reduction treatments. Instead, it was reground in the ball mill using tungsten carbide media at 600 rpm for 3 min and placed into a cylindrical (diameter = 6 mm) vanadium can and time-of-flight ND data were collected at the POWGEN beamline at the spallation neutron source at Oak Ridge National Laboratory. All data were collected at 26.85 °C with a central wavelength of 1.066 Å. Another portion of this sample was packed into a cylindrical Kapton capillary (inner diameter = 0.8 mm) and both ends were sealed with Q Compound (Apiezone). Synchrotron XRD data were collected at 11-BM on the advanced photon source at Argonne National Laboratory with a wavelength of 0.459 Å at 26.85 °C. Structural determinations used these data in joint Rietveld refinements via EXPGUI using GSAS [31,32]. One of the six pellets of each composition was removed after firing and a second pellet was removed after the reduction. Each pellet was ground in an agate mortar and pestle with ~10% silicon by mass, packed onto a flat glass slide, and subjected to Cu radiation in an Ultima IV (Rigaku, Tokyo, Japan) X-ray diffractometer. Scans were taken with 2θ beginning at 10 and ending at 70. The non-reduced patterns were the input used to calculate lattice parameters via whole pattern fitting (MDI Jade 2010). Silicon was used as an internal standard to correct for instrumental offset.

Conductivity studies were performed with a four-point probe. The sheet resistance of each pellet was measured after firing and reduction at five different locations on each face at room temperature using a four-point probe (Model 280PI, Four Dimensions, Inc. Hayward, CA, USA). Sample dimensions and masses were taken at each step to apply geometry and thickness corrections and the sheet resistance was converted to bulk conductivity [33]. Additionally, a porosity correction following the Bruggeman symmetric medium model was applied to each result, assuming that the air in the pellet acts as an insulating phase, following the same procedure as reported previously [13,34].

Band gaps were determined via diffuse reflectance. The diffuse reflectance of each pellet was measured from 250 to 800 nm at room temperature using a Lambda 1050 UV-Vis spectrophotometer with an integrating sphere attachment (PerkinElmer, Inc. Waltham, MA, USA) after firing and reduction. Background spectra were performed on compacted polytetrafluoroethylene. Optical band gaps were calculated by converting diffuse reflectance into Kubelka-Munk notation and approximating the optical band gap as the intersection of a linear extrapolation of the band edge and a linear extrapolation of the background [35–37].

Seebeck coefficients were measured on a Seebeck Measurement System (MMR Technologies, San Jose, CA, USA). Pre-and post-reduced samples were prepared as cylindrical pellets as described above. These were then cut down to ~1 mm × ~1 mm × ~5 mm rectangular prisms and mounted on an alumina stage with silver paste. Five measurements were taken every 15 °C between 305 K and 590 K.

4.3. $Zn_xIn_{2-2x}Sn_xO_3$ Computational Procedure

The structural energies and electronic properties were calculated using spin restricted DFT employed through the Vienna ab initio simulation package (VASP) [38–41]. Plane augmented wave (PAW) pseudopotentials were used with Perdew-Becke-Ernzerhof (PBE) exchange-correlation functionals [42]. A 2 × 2 × 2 k-point grid and an energy cutoff of 400 eV were used in initial structural relaxations. The final grid was augmented (up to 5 × 5 × 5) to guarantee adequate energy precision to distinguish between differing site configurations. Full 80 atom bixbyite unit cells were used, with varying amounts of pairwise zinc and tin substitution on the indium sites. To reduce computational costs, structural relaxations were mainly performed with tin and indium *d*-electrons in the pseudopotential core. Static calculations on relaxed cells were performed with indium *d*-electrons in the valence space to obtain possibly more accurate total energy values, and to assess possible effects on spectroscopic properties. Zinc *d*-states were always treated as valence, as the Zn *d*-band falls only a few eV below the Fermi energy. Although quantitative energies changed somewhat, energy ordering with respect to doping and ordering of structures was essentially unaltered. In retrospect, it was found that inclusion of the nominally fully occupied d^{10} shells of In, Sn and Zn in the valence space was useful in interpreting covalent contributions to bonding, and in elucidating subtle features of the optical densities of states.

5. Conclusions

In summary, we performed a thorough computational examination of $Zn_xIn_{2-2x}Sn_xO_3$ and an experimental examination of $Mg_xIn_{2-2x}Sn_xO_3$ to determine site preferences of the substituting atoms. In both instances, and the prior reported materials $M_xIn_{2-2x}Sn_xO_3$ (M = Ni, Cu, or Zn), a M^{2+} presence on the *d*-site of bixbyite and a Sn^{4+} presence on the *b*-site are suggested. Furthermore, the formation energies of $Zn_xIn_{2-2x}Sn_xO_3$ suggest a short range neighbor pairing of Zn and Sn. Property characterizations of $Mg_xIn_{2-2x}Sn_xO_3$ demonstrate that a reduction treatment makes its conductivity and optical properties much more competitive as a transparent conductor while continuing to be n-type despite the counterbalancing effect of Mg^{2+}.

Supplementary Materials: The following are available online at www.mdpi.com/2073-4352/7/2/47/s1, Figure S1: Lattice Parameters of $Mg_xIn_{2-2x}Sn_xO_{16}$ and $Zn_xIn_{2-2x}Sn_xO_{16}$, Figure S2: XRD patterns of $Mg_xIn_{2-2x}Sn_xO_{16}$, Figure S3: Rietveld Refinement Pattern for $Mg_{0.1}In_{1.8}Sn_{0.1}O_{16}$, Figure S4: Seebeck Coefficients of $Mg_xIn_{2-2x}Sn_2O_{16}$.

Acknowledgments: Karl Rickert recognizes that this material is based upon work supported by the National Science Foundation Graduate Research Fellowship Program under Grant No. DGE-1324585. Any opinions, findings, and conclusions or recommendations expressed in this material are those of the author(s) and do not necessarily reflect the views of the National Science Foundation. N.S. gratefully acknowledge that this study was partially supported by the Council of Higher Education (CoHE) of Turkey. Karl Rickert, Nazmi Sedefoglu, Jeremy Harris, and Kenneth R. Poeppelmeier gratefully acknowledge additional support from the Department of Energy Basic Energy Sciences Award No. DE-FG02-08ER46536. A portion of this research was performed at Oak Ridge National Laboratory's Spallation Neutron Source at POWGEN, which is sponsored by the Scientific User Facilities Division, Office of Basic Energy Sciences, and the U.S. Department of Energy. Use of 11BM at the Advanced Photon Source at Argonne National Laboratory was supported by the U.S. Department of Energy, Office of Science, Office of Basic Energy Sciences, under Contract No. DE-AC02-06CH11357. This work made use of the J. B. Cohen X-ray Diffraction Facility which is supported by the MRSEC program of the National Science Foundation (DMR-1121262) at the Materials Research Center of Northwestern University. A portion of this work was supported by the NU Keck Biophysics Facility and a Cancer Center Support Grant (NCI CA060553).

Author Contributions: Karl Rickert, Nazmi Sedefoglu, Hamide Kavak, and Kenneth R. Poeppelmeier conceived and designed the experiments for $Mg_xIn_{2-2x}Sn_xO_3$. Jeremy Harris and Donald E. Ellis conceived and designed the experiments for $Zn_xIn_{2-2x}Sn_xO_3$. Karl Rickert, Jeremy Harris, and Nazmi Sedefoglu performed the experiments for their respective materials. Karl Rickert wrote the paper and analyzed the data.

Conflicts of Interest: The authors declare no conflict of interest.

References

1. Minami, T. Transparent conducting oxide semiconductors for transparent electrodes. *Semicond. Sci. Technol.* **2005**, *20*, S35–S44. [CrossRef]
2. Granqvist, C.G. Transparent conductors as solar energy materials: A panoramic review. *Sol. Energy Mater. Sol. Cells* **2007**, *91*, 1529–1598. [CrossRef]
3. Ginley, D.S.; Perkins, J.D. Transparent Conductors. In *Handbook of Transparent Conductors*; Ginley, D.S., Ed.; Springer: Boston, MA, USA, 2011; pp. 1–25.
4. Frank, G.; Kostlin, H. Electrical-properties and defect model of tin-doped indium oxide layers. *Appl. Phys. A* **1982**, *27*, 197–206. [CrossRef]
5. Bizo, L.; Choisnet, J.; Retoux, R.; Raveau, B. The great potential of coupled substitutions in In_2O_3 for the generation of bixbyite-type transparent conducting oxides, $In_{2-2x}M_xSn_xO_3$. *Solid State Commun.* **2005**, *136*, 163–168. [CrossRef]
6. Hoel, C.A.; Mason, T.O.; Gaillard, J.-F.; Poeppelmeier, K.R. Transparent conducting oxides in the $ZnO-In_2O_3-SnO_2$ system. *Chem. Mater.* **2010**, *22*, 3569–3579. [CrossRef]
7. Harvey, S.P.; Mason, T.O.; Buchholz, D.B.; Chang, R.P.H.; Korber, C.; Klein, A. Carrier generation and inherent off-stoichiometry in Zn, Sn codoped indium oxide (ZITO) bulk and thin-film specimens. *J. Am. Ceram. Soc.* **2008**, *91*, 467–472. [CrossRef]
8. Nomura, K.; Ujihira, Y.; Tanaka, S.; Matsumoto, K. Characterization and estimation of ito (indium-tin-oxide) by mossbauer spectrometry. *Hyperfine Interact.* **1988**, *42*, 1207–1210. [CrossRef]
9. Nadaud, N.; Lequeux, N.; Nanot, M.; Jové, J.; Roisnel, T. Structural studies of tin-doped indium oxide (ITO) and $In_4Sn_3O_{12}$. *J. Solid State Chem.* **1998**, *135*, 140–148. [CrossRef]
10. Mryasov, O.N.; Freeman, A.J. Electronic band structure of indium tin oxide and criteria for transparent conducting behavior. *Phys. Rev. B* **2001**, *64*, 233111. [CrossRef]
11. Palmer, G.B.; Poeppelmeier, K.R.; Mason, T.O. Conductivity and transparency of ZnO/SnO_2-cosubstituted In_2O_3. *Chem. Mater.* **1997**, *9*, 3121–3126. [CrossRef]
12. Lu, Y.-B.; Yang, T.L.; Ling, Z.C.; Cong, W.-Y.; Zhang, P.; Li, Y.H.; Xin, Y.Q. How does the multiple constituent affect the carrier generation and charge transport in multicomponent tcos of In-Zn-Sn oxide. *J. Mater. Chem. C* **2015**, *3*, 7727–7737. [CrossRef]
13. Rickert, K.; Sedefoglu, N.; Malo, S.; Caignaert, V.; Kavak, H.; Poeppelmeier, K.R. Structural, electrical, and optical properties of the tetragonal, fluorite-related $Zn_{0.456}In_{1.084}Ge_{0.460}O_3$. *Chem. Mater.* **2015**, *27*, 5072–5079. [CrossRef]
14. Proffit, D.E.; Buchholz, D.B.; Chang, R.P.H.; Bedzyk, M.J.; Mason, T.O.; Ma, Q. X-ray absorption spectroscopy study of the local structures of crystalline Zn-In-Sn oxide thin films. *J. Appl. Phys.* **2009**, *106*, 113524. [CrossRef]
15. Hoel, C.A.; Gaillard, J.F.; Poeppelmeier, K.R. Probing the local structure of crystalline ZITO: $In_{2-2x}Sn_xZn_xO_3$ ($x \leq 0.4$). *J. Solid State Chem.* **2010**, *183*, 761–768. [CrossRef]
16. Hoel, C.A.; Gallardo Amores, J.M.; Moran, E.; Angel Alario-Franco, M.; Gaillard, J.-F.; Poeppelmeier, K.R. High-pressure synthesis and local structure of corundum-type $In_{2-2x}M_xSn_xO_3$ ($x \leq 0.7$). *JACS* **2010**, *132*, 16479–16487.
17. Hoel, C.A.; Xie, S.; Benmore, C.; Malliakas, C.D.; Gaillard, J.-F.; Poeppelmeier, K.R. Evidence for tetrahedral zinc in amorphous $In_{2-2x}Zn_xSn_xO_3$ (a-ZITO). *Z. Anorg. Allg. Chem.* **2011**, *637*, 885–894. [CrossRef]
18. Hoel, C.A.; Buchholz, D.B.; Chang, R.P.H.; Poeppelmeier, K.R. Pulsed-laser deposition of heteroepitaxial corundum-type zito: $Cor-In_{2-2x}Zn_xSn_xO_3$. *Thin Solid Films* **2012**, *520*, 2938–2942. [CrossRef]
19. Proffit, D.E.; Philippe, T.; Emery, J.D.; Ma, Q.; Buchholz, B.D.; Voorhees, P.W.; Bedzyk, M.J.; Chang, R.P.H.; Mason, T.O. Thermal stability of amorphous Zn-In-Sn-O films. *J. Electroceram.* **2015**, *34*, 167–174. [CrossRef]
20. Ni, J.; Wang, L.; Yang, Y.; Yan, H.; Jin, S.; Marks, T.J.; Ireland, J.R.; Kannewurf, C.R. Charge transport and optical properties of mocvd-derived highly transparent and conductive Mg- and Sn-doped In_2O_3 thin films. *Inorg. Chem.* **2005**, *44*, 6071–6076. [CrossRef] [PubMed]
21. Vegard, L. The constitution of the mixed crystals and the filling of space of the atoms. *Z. Phys.* **1921**, *5*, 17–26. [CrossRef]

22. Shannon, R.D. Revised effective ionic-radii and systematic studies of interatomic distances in halides and chalcogenides. *Acta Crystallogr. Sect. A Found. Crystallogr.* **1976**, *32*, 751–767. [CrossRef]

23. Denton, A.R.; Ashcroft, N.W. Vegard law. *Phys. Rev. A* **1991**, *43*, 3161–3164. [CrossRef] [PubMed]

24. Haas, P.; Tran, F.; Blaha, P. Calculation of the lattice constant of solids with semilocal functionals. *Phys. Rev. B* **2009**, *79*, 085104. [CrossRef]

25. Sæterli, R.; Flage-Larsen, E.; Friis, J.; Løvvik, O.M.; Pacaud, J.; Marthinsen, K.; Holmestad, R. Experimental and theoretical study of electron density and structure factors in CoSb$_3$. *Ultramicroscopy* **2011**, *111*, 847–853. [CrossRef] [PubMed]

26. Hamilton, W.C. Significance tests on crystallographic r factor. *Acta Crystallogr.* **1965**, *18*, 502–510. [CrossRef]

27. Sunde, T.O.L.; Lindgren, M.; Mason, T.O.; Einarsrud, M.A.; Grande, T. Solid solubility of rare earth elements (Nd, Eu, Tb) in In$_{2-x}$Sn$_x$O$_3$—Effect on electrical conductivity and optical properties. *Dalton Trans.* **2014**, *43*, 9620–9632. [CrossRef] [PubMed]

28. Burstein, E. Anomalous optical absorption limit in InSb. *Phys. Rev.* **1954**, *93*, 632–633. [CrossRef]

29. Moss, T.S. The interpretation of the properties of indium antimonide. *Proc. Phys. Soc. Lond. Sect. B* **1954**, *67*, 775–782. [CrossRef]

30. Bizo, L.; Choisnet, J.; Raveau, B. Coupled substitutions in In$_2$O$_3$: New transparent conductors In$_{2-2x}$M$_{2x/3}$Sb$_{x/3}$O$_3$ (M = Cu, Zn). *Mater. Res. Bull.* **2006**, *41*, 2232–2237. [CrossRef]

31. Larson, A.C.; Von Dreele, R.B. *General structure analysis system (GSAS)*; Report LAUR; Los Alamos National Laboratory: Los Alamos, NM, USA; pp. 86–748.

32. Toby, B.H. 'Expgui', a graphical user interface for gsas. *J. Appl. Crystallogr.* **2001**, *34*, 210–213. [CrossRef]

33. Smits, F.M. Measurement of sheet resistivities with the 4-point probe. *Bell Syst. Tech. J.* **1958**, *37*, 711–718. [CrossRef]

34. McLachlan, D.S.; Blaszkiewicz, M.; Newnham, R.E. Electrical-resistivity of composites. *J. Am. Ceram. Soc.* **1990**, *73*, 2187–2203. [CrossRef]

35. Kubelka, P.; Munk, Z. An article on optics of paint layers. *Z. Tech. Phys.* **1931**, *12*, 593–603.

36. Wendlandt, W.W.; Hecht, H.G. *Reflectance Spectroscopy*; Interscience Publishers: New York, NY, USA, 1966.

37. Tandon, S.P.; Gupta, J.P. Measurement of forbidden energy gap of semiconductors by diffuse refectance technique. *Phys. Status Solidi* **1970**, *38*, 363. [CrossRef]

38. Kresse, G.; Hafner, J. Ab initio molecular-dynamics for liquid-metals. *Phys. Rev. B* **1993**, *47*, 558–561. [CrossRef]

39. Kresse, G.; Hafner, J. Ab-initio molecular-dynamics simulation of the liquid metal amorphous-semiconductor transition in germanium. *Phys. Rev. B* **1994**, *49*, 14251–14269. [CrossRef]

40. Kresse, G.; Furthmuller, J. Efficient iterative schemes for ab initio total-energy calculations using a plane-wave basis set. *Phys. Rev. B* **1996**, *54*, 11169–11186. [CrossRef]

41. Kresse, G.; Furthmuller, J. Efficiency of ab-initio total energy calculations for metals and semiconductors using a plane-wave basis set. *Comput. Mater. Sci.* **1996**, *6*, 15–50. [CrossRef]

42. Perdew, J.P.; Burke, K.; Ernzerhof, M. Generalized gradient approximation made simple. *Phys. Rev. Lett.* **1996**, *77*, 3865–3868. [CrossRef] [PubMed]

© 2017 by the authors. Licensee MDPI, Basel, Switzerland. This article is an open access article distributed under the terms and conditions of the Creative Commons Attribution (CC BY) license (http://creativecommons.org/licenses/by/4.0/).

crystals

MDPI

Article

Epoxy-Based Composites Embedded with High Performance BZT-0.5BCT Piezoelectric Nanoparticles Powders for Damping Vibration Absorber Application

Zengmei Wang [1,*], Huanhuan Wang [1], Wenyan Zhao [1] and Hideo Kimura [2,*]

[1]   School of Materials Science and Engineering, Southeast University, Nanjing 211189, China; fengxinzi.huan@hotmail.com (H.W.); 18651620827@163.com (W.Z.)
[2]   National Institute for Materials Science, Tsukuba 305-0047, Japan
*   Correspondence: 101011338@seu.edu.cn (Z.W.); kimura.hideo@nims.go.jp (H.K.); Tel.: +86-255-209-1085 (Z.W.); +81-29-859-2437 (H.K.)

Academic Editor: Stevin Snellius Pramana
Received: 24 February 2017; Accepted: 6 April 2017; Published: 9 April 2017

Abstract: Lead-free, high piezoelectric performance, $Ba(Zr_{0.2}Ti_{0.8})O_3$-$0.5(Ba_{0.7}Ca_{0.3})TiO_3$ (BZT-0.5BCT) sub-micron powders with perovskite structure were fabricated using the sol-gel process. A 0–3 type composite was obtained by choosing epoxy resin as matrix and BZT-0.5BCT, acetylene black as functional phases. Particular attention was paid to the damping behavior of composite with different content of BZT-0.5BCT powders, the influence of frequency and loading force on the damping properties were also analyzed. A mathematical model was developed to characterize the damping properties of the composites. It found that the piezoelectric effects and interfacial friction play a key role in damping behavior of composites, and a large dissipated loss factor of tanδ was found at the BZT-0.5BCT content of 20 vol %.

Keywords: lead-free piezoelectric composites; vibration absorber; damping properties

---

1. Introduction

Unwanted vibrations may result in the fatigue and failure of structures from concrete steel buildings to precision instruments; therefore, vibration reduction and noise control is a serious engineering challenge. Much effort has been expended in attempting to minimize mechanical vibrations [1–4]. One of the approaches is to use damping materials to limit the vibration [5,6]. Viscoelastic material damping is recognized to make a contribution to energy loss due to hysteresis caused by internal sliding of the molecule chain. Thus, it is expected that the polymer possesses a broad, high damping peak applied to damping applications [7]. However, this is difficult or impossible for a single polymer because the height and width of the loss peak cannot be independently adjusted. Composites with piezoelectric materials contained showed excellent damping properties with wide temperature range and have attracted ever-greater attention [8,9].

Piezoelectric materials such as lead zirconate titanate (PZT), lead magnesio niobate (PMN), barium titanate (BT), zinc oxide (ZnO), poly(vinylidene fluoride) (PVDF), etc., have been intensively investigated for great ability of converting mechanical energy into electricity [10–14]. The electric energy can be dissipated through an external resistance in piezo-damping composites [15]. Hori et al. exploited a damper material which is composed of PZT powders, conductive carbon (CB) powders and epoxy (EP) resin, and studied their damping properties and the effects of CB content on the loss factor [16]. Tang et al. prepared a series of $BaTiO_3$ filled polyurethane (PU)/unsaturated polyester resin (UP) interpenetrating polymer networks (IPNs) and explored the relationship between dielectric properties and damping properties of the mixtures [17]. Malakooti et al. fabricated vertically aligned ZnO nanowires on nanofillers and produced hybrid carbon fiber composites. The results showed that the morphology-controlled interphase between reinforcement and matrix not only exhibited remarkable damping enhancement but also stiffness improvement [18]. Recently, a lead-free $Ba(Zr_{0.2}Ti_{0.8})O_3$-$x(Ba_{0.7}Ca_{0.3})TiO_3$ (BZT-xBCT) piezoelectric system with optimal composition of x = 0.5 was reported to show outstanding room temperature piezoelectricity with the piezoelectric coefficient d_{33} = 560–620 pC/N which is comparable to PZT [19]. Zhou, et al. developed a scalable method to synthesize perovskite type $0.5Ba(Zr_{0.2}Ti_{0.8})O_3$–$0.5(Ba_{0.7}Ca_{0.3})TiO_3$ (BZT–BCT) NWs with high piezoelectric coupling coefficient (90 pm/V) and the energy harvesting performance of the BZT–BCT NWs were demonstrated by fabricating a flexible nanocomposite exhibiting a high power density (2.25 μW/cm)[20]. Wu et al. synthesized a new kind of lead-free $0.5Ba(Zr_{0.2}Ti_{0.8})O_3$–$0.5(Ba_{0.7}Ca_{0.3})TiO_3$ (BZT–BCT) NWs with high piezoelectric coefficients through the electrospinning method and subsequent calcining process which could harvest weak mechanical movement energy and generated an output voltage of 3.25 V and output current of 55 nA [21]. Therefore, there is much promise in such a high-performance lead-free piezoelectric material used in polymer-based damping materials through forming the composites containing piezoelectric phase and polymer matrix.

In this paper, we added BZT-0.5BCT powders and acetylene black powders into epoxy resin, and fabricated a series of piezo-damping composites through solution blending. In line with the corresponding ceramic counterpart, BZT-0.5BCT shows the highest value of Pr (22.15 mC/cm²) and fairly low Ec (68.06 kV/cm) [22]. The effects of different contents of BZT-0.5BCT, frequency and loading force on damping properties were investigated. A mathematical model was established to understand the piezoelectric damping mechanism in the composites.

2. Results and Discussions

Figure 1 shows the XRD patterns of the three BZT-0.5BCT/acetylene black/epoxy composites. It can be observed that without addition of piezoelectric phase, Epoxy E51 shown amorphous phase, with addition of 20 vol %, 40 vol % and 60 vol % piezoelectric material. All samples show the polycrystalline perovskite phase.

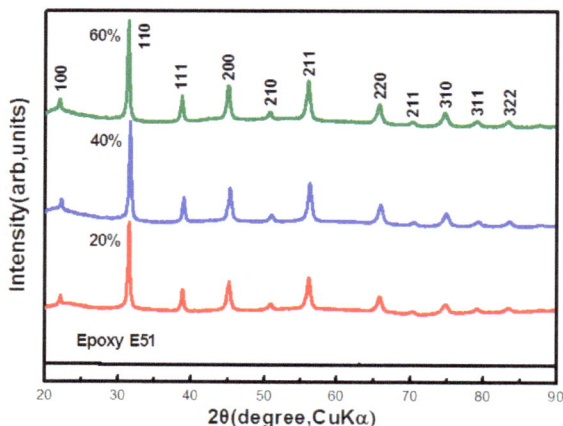

Figure 1. X-ray diffractometer (XRD) patterns of composites with different content of piezoelectric phase.

The particle distribution of the BZT-0.5BCT powder incorporated in the 0–3 epoxy-based piezoelectric composites was measured with sub-micron particle size analysis. According to the cumulative curve given in Figure 2a, the average grain size was around 500 nm. Figure 2b–d present the typical SEM image of the cross-section of the composite with 20 vol %, 40 vol % and 60 vol %

powders contained, respectively. It can be clearly observed that with the increase of piezoelectric phase, most oval BZT-0.5BCT particles and conductive acetylene black were uniformly dispersed in the matrix of Epoxy resin. A small amount of reunion, the grain size of piezoelectric phase was consistent with that in Figure 2a. However, when the volume fraction reached up to 60%, the nanoparticles tended to agglomerate due to too high content of the piezoelectric phase.

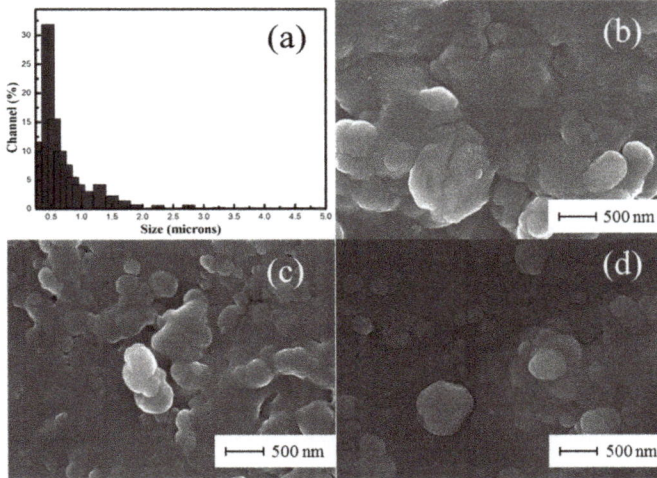

Figure 2. Typical particle size distribution histogram of $Ba(Zr_{0.2}Ti_{0.8})O_3$-$0.5(Ba_{0.7}Ca_{0.3})TiO_3$ (BZT-0.5BCT) powders (**a**); Typical SEM image of composite with 20 vol % (**b**), 40 vol % (**c**) and 60 vol % (**d**) BZT-0.5BCT powders contained.

DMA tests generally characterize the damping properties of materials such as the storage modulus (E′) as measurement of dynamic stiffness, loss modulus (E″) as a measurement for energy dissipation and the tangent of phase angle between the stress and strain vectors linking these two moduli (tan δ). Damping loss factor reflects an ability of converting the mechanical energy into heat energy when the material is subjected to external cyclic loading. Figure 3 shows the effect of different BZT-0.5BCT content on the loss factor of the composites when the preload fixed at 0.5 N and frequency 10 Hz. In the vicinity of the glass transition temperature (Tg) of the composites, the peaks shifted with variation of composing component.

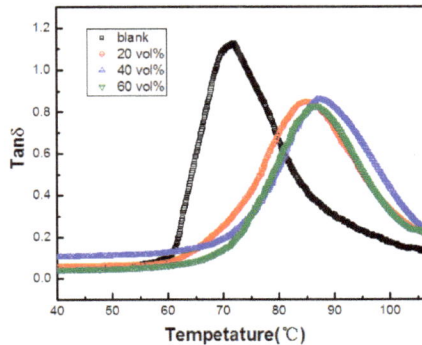

Figure 3. Loss factor-temperature curves of composite with different BZT-0.5BCT content.

Detailed damping properties are shown in Table 1. It indicated when different volume fraction BZT-0.5BCT (20%, 40% and 60%) was added to epoxy resin, the maximum loss factor and the effective temperature region initially increase and then decrease as the increase of the piezoelectric phase within our limited test range. The damping property of the composites resulted from comprehensive function of interfacial friction dissipation, retarded motion of polymer macromolecular and the piezoelectric effect. Because of the piezoelectric effect of BZT-0.5BCT powder, mechanical energy changes into electrical energy when the composites were subjected to external vibration. During the glass transition region, the composite chain movement happened under the impact load, and the piezoelectric effect of the piezoelectric phase consumed a strong response to the load which resulted in the maximum hysteresis damping and maximum lost energy. When BZT-0.5BCT powders content ranges from 20 to 40 vol %, the interaction between the piezoelectric phase and epoxy matrix became stronger and limitation for piezoelectric powder to the polymeric segmental relaxation would enhanced. Meanwhile, T_g grew from 71.8 °C to 87.5 °C as a result and widened the effective temperature range. However, when the volume fraction of the nanoparticles is 60%, a large number of two-phase interfaces formed and the epoxy resin molecules in the vicinity of the interface layer was blocked from moving, which led to the piezoelectric effect not being able to be fully utilized and a reduction of the dissipated energy. Only when the volume fraction of BZT-0.5BCT was 0.2 can the combined effects contribute to the most powerful enhancement of damping properties. The maximum loss factor was 0.846 when the effective temperature range was maximum.

Table 1. Damping Properties of Composites with Different BZT-0.5BCT Powders Content.

Volume Fraction (vol %)	Maximum Loss Factor (tan δ_{max})	Glass Transition Temperature Tg (°C)	Temperature Range (tan δ > 0.3) (°C)
0	1.135	71.8	62.6~90.1(27.5)
20	0.846	84.9	71.7~101.3(29.6)
40	0.857	87.5	74.8~103.9(29.1)
60	0.826	86.7	75.1~100.7(25.6)

Figure 4 and Table 2 show the damping properties of composite with 40 vol % BZT-0.5BCT powders under multi-frequency (1, 2, 4, 10 and 20 Hz). The maximum loss factor, glass transition temperature and the effective temperature range were all improved when the frequency gets higher and there was a trend of stability. The higher frequency speeded up the molecular chain movement and enhanced the piezoelectric activity of the BZT-0.5BCT powders, thus causing the improvement in the damping properties and the right shift of the peaks in the tan δ curves.

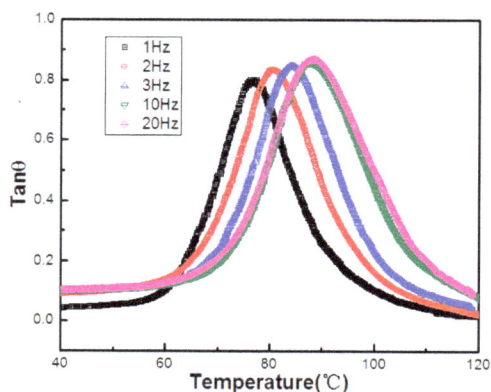

Figure 4. Loss factor-temperature curves of composites at multi-frequency.

Table 2. Damping Properties of Composite at Different Frequency.

Frequency (Hz)	Maximum Loss Factor (tan δ_{max})	Glass Transition Temperature Tg (°C)	temperature Range (tan $\delta > 0.3$) (°C)
1	0.803	76.6	66.3~88.5(22.2)
2	0.835	80.3	68.8~94.2(25.4)
4	0.848	84.0	71.9~97.7(25.8)
10	0.858	87.4	74.8~103.8(29.0)
20	0.869	88.2	74.1~105.7(31.6)

Figure 5 and Table 3 show the effects of loading force of composite with 20 vol % addition of BZT-0.5BCT powders with the loading force ranging from 0.5 N to 4 N. The maximum damping factor increased when the loading force was raised. Meanwhile, T_g increased at first then remained more or less the same. The effective temperature range of the composite showed almost no change, and reached a peak at the loading force of 4 N.

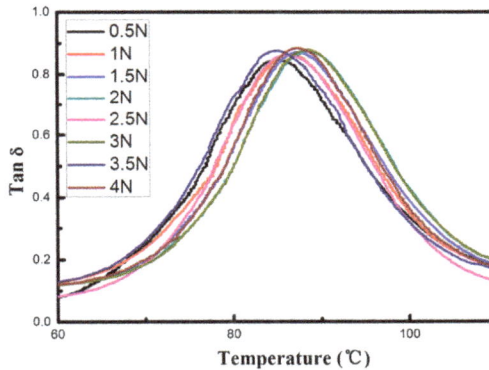

Figure 5. Loss factor-temperature curves of composites with different loading force.

Table 3. Damping Properties of Composite at Different Loading Force.

Loading Force (N)	Maximum Loss Factor (tan δ)	Glass Transition Temperature Tg (°C)	Temperature Range (tan $\delta > 0.3$) (°C)
0.5	0.847	84.9	71.7~101.3 (29.6)
1	0.863	86.4	72.4~101.6 (29.2)
1.5	0.871	87.3	74.1~102.9 (28.8)
2	0.874	88.6	75.1~103.7 (28.6)
2.5	0.876	88.5	75.0~103.7 (28.7)
3	0.879	88.5	75.0~103.7 (28.7)
3.5	0.875	85.0	71.3~100.6 (29.3)
4	0.884	87.3	74.3~101.9 (27.6)

The composites behavior shown with different loading force is not like that of a conventional viscoelastic body [23]. When samples were with the preload force, the bond length and bond-angle of epoxy polymer's molecular chain would change as a tensioning process, which would cause an obstacle to the deformation of the composites. The deformable parts in the matrix become smaller while the preload force increases. On the other hand, when the preload force raises, the piezoelectric effect and the interface friction would enhance, causing an increase in the maximum loss factor. T_g and effective temperature region mostly depended on the chain segment movement of epoxy so the peaks in tanδ curves shift slightly to the right side. We can speculate that the piezoelectric effect and interfacial friction mainly contribute to composites' damping properties when the piezoelectric ceramic volume fraction exceeds 0.2. The damping capacity was enhanced by piezo-damping as the piezoelectric effect enlarged in the piezoelectric ceramic when the loading force increased. However, the changing range was lower.

To understand the damping mechanism in the piezoelectric/polymer composites, we can analyze the components and electric circuit in the piezoelectric composite. The damping function of the piezoelectric composite mainly originates from three parts: the viscoelastic damping of the epoxy resin, the interface friction damping between the matrix and the fillers, and piezoelectric damping of BZT-0.5BCT powders. As for the piezoelectric damping, the ceramic powders first generate electric energy upon deformation and then dissipate them as thermal energy through the conductive matrix. For the piezoelectric/polymer composites, we can assume that each piezoelectric particle was in contact with the matrix as shown in Figure 6, and could be equivalent to a RLC circuit in a parallel connection. Figure 7 shows a mathematical model of the BZT-0.5BCT/acetylene black/epoxy composite with 0–3 connectivity that is the equivalent circuit between them. Then an adjustable resistance was used to simulate the conductivity of the polymer matrix and we supposed there were n piezoelectric particles.

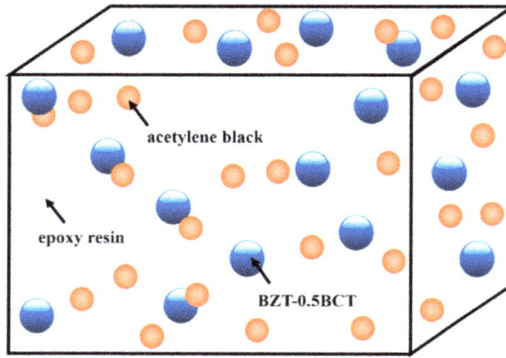

Figure 6. Schematic representation of composite.

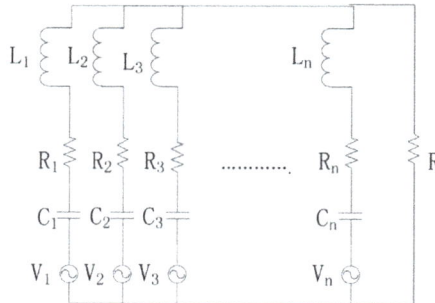

Figure 7. Equivalent circuit of composite.

When a force is applied to the composite, in a single circuit unit an alternating voltage V_n produced by piezoelectric powder is generated and the electric energy is dissipated by the resistance R. The impedance (Z) in the circuit cell is

$$Z_n = R + \cfrac{1}{j\omega L_n + \frac{1}{j\omega C_n} + R_n} \qquad (1)$$

the consumption energy (W_n) by R is expressed as

$$W_n = \frac{V_n^2 \cdot R}{\left[R + \frac{R_n}{R_n^2 + \left(\omega L - \frac{1}{\omega C}\right)^2}\right]^2 + \left[\frac{\omega L - \frac{1}{\omega C}}{R_n^2 + \left(\omega L - \frac{1}{\omega C}\right)^2}\right]^2} \qquad (2)$$

When it works at a resonance frequency,

$$L = \frac{1}{\omega C} \qquad (3)$$

W_n takes a maximum value in Equation (2), that is

$$W_n = \frac{V_n^2 \cdot R}{\left(R + \frac{1}{R_n}\right)^2} \qquad (4)$$

For the piezoelectric effect of piezoelectric materials, the mechanical energy transferred into electrical energy. Then it conversed into heat consumption through the conductive pathways and the load in composite materials. This indicated that the piezoelectric damping of composite was in closer association with the content of BZT-0.5BCT powders, frequency and the force, as the piezoelectric phase content directly affects the value of n and the loading force has an influence on the voltage (V_n).

3. Materials and Methods

BZT-0.5BCT nanopowders were prepared by a sol-gel process. The starting reagents were analytical graded barium acetate [$Ba(CH_3COO)_2$], calcium acetate monohydrate [$Ca(CH_3COO)_2 \cdot H_2O$], tetrabutyl titanate [$Ti(OC_4H_9)_4$], zirconium oxynitrate [$ZrO(NO_3) \cdot 2H_2O$], complexing agent of glacial acetic acid and solvent of 2-methoxyethanol. After obtaining sol solution, the dry gel was ground and followed post heat-treatment, then piezoelectric sub-micron powders were obtained.

The epoxy resin matrix (E-51) used in this research was provided by Wuxi Xinmeng Co. Ltd. (Wuxi, China) and acetylene black powders were provided by Taiyuan Yingze battery sales department (Taiyuan, China). To improve the surface lipophilic, BZT-0.5BCT powders were firstly treated by silane coupling agents and absolute alcohol. Conductive acetylene black was added into epoxy resin and mixed, then the pretreated piezoelectric powders and curing agents were added and stirred thoroughly under vacuum. The homogeneous viscous mixture was poured into a mold and solidified at 80 °C for 2 h. The BZT-0.5BCT/acetylene black/epoxy composite was obtained when cuing finished. The composite was cut, and poled. The three different composites fabricated contain 2 wt % acetylene black and 20, 40 and 60 vol % piezoelectric powders, respectively. Five 20 mm × 3 mm × 2 mm samples were fabricated for dynamic mechanical analysis.

The phase and morphology observations of composites were examined by X-ray diffractometer (XRD, Bruker D8 Discover) with Cu-Kα (1.5406 Å), tube voltage (40 kV), tube current (100 mA), scanning speed (0.15 degrees/s) and Field emission scanning electron microscope (FESEM, FEI Sirion-200). The damping properties of materials represented by the value of loss factor (tanδ) were carried out using Dynamic Mechanical Analyzer (DMA, TA-Q800). The measurements were made at different frequency and loading force in the temperature range of 0~120 °C at a heating rate of 2 °C/min with multi-frequency-strain mode.

4. Conclusions

In this work, 0–3 BZT-0.5BCT/ acetylene black/epoxy composites with different volume fraction ratio were prepared via sol-gel process and pouring method. The microstructure and morphology of composites were studied and the damping properties of the composite with different volume fraction, frequency and loading force were mainly related. The addition of piezoelectric phase intrinsically broadens its effective temperature region from 27.52 °C to 29.56 °C with only 20 vol % BZT-0.5BCT powders added and spurred the glass transition shift to a high temperature section. The composite damping ratio is a sum of the polymer damping and piezo-damping. The piezo effect and interfacial friction play a major role in damping capacity of the piezoelectric-damping composites when the piezo-ceramic volume fraction exceeds 0.2. The maximum loss factor rose as the content of BZT-0.5BCT powders, frequency and loading force reached a threshold level. The composites show less sensitivity to stress due to the fact that the BZT-0.5BCT powders can retain their piezoelectric properties in comparison to polymer damping materials. Also, an equivalent circuit was established to understand the mechanism of the piezoelectric damping properties.

Acknowledgments: This work is financially sponsored by Natural Science Foundation of China (Grants No. 51002029). Part of this work was supported by the GRENE (Green Network of Excellence) project by Ministry of Education, Culture, Sports, Science and Technology-Japan.

Author Contributions: Zengmei Wang and HuanHuan Wang conceived and designed the experiments; Huanhuan Wang performed the experiments; Wenyan Zhao and Hideo Kimura analyzed the data; Wenyan Zhao revised the paper.

Conflicts of Interest: The founding sponsors had no role in the design of the study; in the collection, analyses, or interpretation of data; in the writing of the manuscript, and in the decision to publish the results.

References

1. Amick, H.; Gendreau, M. Construction vibrations and their impact on vibration-sensitive facilities. *ASCE Constr. Congr.* **2000**. [CrossRef]
2. Yu, X.; Fu, Y. Dynamic analysis of damage behavior of vibration compaction on concrete bridges. In Proceedings of the Seventh International Conference on Traffic and Transportation Studies, Kunming, China, 3–5 August 2010.
3. Alavinasab, A.; Padewski, E.; Holley, M.; Jha, R.; Ahmadi, G. Damage Identification Based on Vibration Response of Prestressed Concrete Pipes. In Proceedings of the ASCE Pipelines Conference: Climbing New Peaks to Infrastructure Reliability—Renew, Rehab, and Reinvest, Keystone, CO, USA, 28 August–1 September 2010.
4. Fan, W.; Qiao, P. Vibration-based damage identification methods: A review and comparative study. *Struct. Health Monit.* **2011**, *10*, 83–111. [CrossRef]
5. Kwak, G.H.; Inoue, K.; Tominaga, Y.; Asai, S.; Sumita, M. Characterization of the vibrational damping loss factor and viscoelastic properties of ethylene–propylene rubbers reinforced with micro-scale fillers. *J. Appl. Polym. Sci.* **2001**, *82*, 3058–3066. [CrossRef]
6. Saravanan, C.; Ganesan, N.; Ramamurti, V. Vibration and damping analysis of multilayered fluid filled cylindrical shells with constrained viscoelastic damping using modal strain energy method. *Comput. Struct.* **2000**, *75*, 395–417. [CrossRef]
7. Skandani, A.A.; Masghouni, N.; Case, S.W.; Leo, D.J.; Al-Haik, M. Enhanced vibration damping of carbon fibers-ZnO nanorods hybrid composites. *Appl. Phys. Lett.* **2012**, *101*, 73–111.
8. Guo, X.J.; Zhang, J.S.; Xu, D.Y.; Xie, X.C.; Sha, F.; Huang, S.F. Effects of PMN Volume Fraction on the Damping Properties of 1–3 Piezoelectric Damping Composites. *Appl. Mech. Mater.* **2014**, *624*, 8–12. [CrossRef]
9. Suhr, J.; Koratkar, N.A.; Ye, D.; Lu, T.M. Damping properties of epoxy films with nanoscale fillers. *J. Intell. Mater. Syst. Struct.* **2006**, *17*, 255–260. [CrossRef]
10. Anton, S.R.; Sodano, H.A. A review of power harvesting using piezoelectric materials (2003–2006). *Smart Mater. Struct.* **2007**, *16*, R1. [CrossRef]
11. Ren, B.; Or, S.W.; Zhang, Y.; Zhang, Q.; Li, X.; Jiao, J.; Wang, W.; Liu, D.; Zhao, X.; Luo, H. Piezoelectric energy harvesting using shear mode 0.71Pb(Mg$_{1/3}$Nb$_{2/3}$)O$_3$-0.29PbTiO$_3$ single crystal cantilever. *Appl. Phys. Lett.* **2010**, *96*, 083–502. [CrossRef]
12. Deng, Z.; Dai, Y.; Chen, W.; Pei, X.; Liao, J. Synthesis and characterization of bowl-like single-crystalline BaTiO$_3$ nanoparticles. *Nanoscale Res. Lett.* **2010**, *7*, 1217–1221. [CrossRef] [PubMed]
13. Wang, X.D.; Song, J.H.; Liu, J.; Wang, Z.L. Direct-current nanogenerator driven by ultrasonic waves. *Science* **2007**, *5821*, 102–105. [CrossRef] [PubMed]

14. Klimiec, E.; Zaraska, W.; Zaraska, K.; Gasiorski, K.P.; Sadowski, T.; Pajda, M. Piezoelectric polymer films as power converters for human powered electronics. *Microelectron Reliab.* **2008**, *48*, 897–901. [CrossRef]

15. Marra, S.P.; Ramesh, K.T.; Douglas, A.S. The mechanical properties of lead-titanate/polymer 0–3 composites. *Compos. Sci. Technol.* **1999**, *59*, 2163–2173. [CrossRef]

16. Hori, M.; Aoki, T.; Ohira, Y.; Yano, S. New type of mechanical damping composites composed of piezoelectric ceramics, carbon black and epoxy resin. *Compos. Part A* **2001**, *32*, 287–290. [CrossRef]

17. Tang, D.; Zhang, J.; Zhou, D.; Zhao, L. Influence of $BaTiO_3$ on damping and dielectric properties of filled polyurethane/unsaturated polyester resin interpenetrating polymer networks. *J. Mater. Sci.* **2005**, *40*, 3339–3345. [CrossRef]

18. Malakooti, M.H.; Hwang, H.S.; Sodano, H.A. Morphology-Controlled ZnO Nanowire Arrays for Tailored Hybrid Composites with High Damping. *ACS Appl. Mat. Interfaces* **2014**, *7*, 332–339. [CrossRef] [PubMed]

19. Cai, Z.L.; Wang, Z.M.; Wang, H.H.; Cheng, Z.X.; Li, B.W.; Guo, X.L.; Kimura, H.; Kasahara, A. An Investigation of the Nanomechanical Properties of $0.5Ba(Ti_{0.8}Zr_{0.2})O_3$–$0.5(Ba_{0.7}Ca_{0.3})TiO_3$ Thin Films. *J. Am. Ceram. Soc.* **2015**, *98*, 114–118. [CrossRef]

20. Zhou, Z.; Bowland, C.C.; Malakooti, M.H.; Tang, H.X.; Sodano, H.A. Lead-free $0.5Ba(Zr_{0.2}Ti_{0.8})O_3$–$0.5(Ba_{0.7}Ca_{0.3})TiO_3$ nanowires for energy harvesting. *RSC Nanoscale* **2016**, *8*, 5098–5105. [CrossRef] [PubMed]

21. Wu, W.W.; Cheng, L.; Bai, S.; Dou, W.; Xu, Q.; Wei, Z.Y.; Qin, Y. Electrospinning lead-free $0.5Ba(Zr_{0.2}Ti_{0.8})O_3$–$0.5(Ba_{0.7}Ca_{0.3})TiO_3$ nanowires and their application in energy harvesting. *J. Mater. Chem. A* **2013**, *1*, 7332–7338. [CrossRef]

22. Wang, Z.M.; Zhao, K.; Guo, X.L.; Sun, W.; Jiang, H.L.; Han, X.Q.; Tao, X.T.; Cheng, Z.X.; Zhao, H.Y.; Kimura, H.; et al. Crystallization, phase evolution and ferroelectric properties of sol–gel-synthesized $Ba(Ti_{0.8}Zr_{0.2})O_3$–$x(Ba_{0.7}Ca_{0.3})TiO_3$ thin films. *J. Mater. Chem. C* **2013**, *1*, 522–530. [CrossRef]

23. Guo, D.; Mao, W.; Qin, Y.; Huang, Z.; Wang, C.; Shen, Q.; Zhang, L. Damping properties of epoxy-based composite embedded with sol-gel derived $Pb(Zr_{0.53}Ti_{0.47})O_3$ thin film annealed at different temperatures. *J. Mater. Sci. Mater. Electron.* **2012**, *23*, 940–944. [CrossRef]

© 2017 by the authors. Licensee MDPI, Basel, Switzerland. This article is an open access article distributed under the terms and conditions of the Creative Commons Attribution (CC BY) license (http://creativecommons.org/licenses/by/4.0/).

![crystals logo] *crystals*

MDPI

Article

Li$_2$HgMS$_4$ (M = Si, Ge, Sn): New Quaternary Diamond-Like Semiconductors for Infrared Laser Frequency Conversion

Kui Wu and Shilie Pan *

Key Laboratory of Functional Materials and Devices for Special Environments of CAS,
Xinjiang Key Laboratory of Electronic Information Materials and Devices,
Xinjiang Technical Institute of Physics & Chemistry of CAS, 40-1 South Beijing Road,
Urumqi 830011, China; wukui@ms.xjb.ac.cn
* Correspondence: slpan@ms.xjb.ac.cn; Tel.: +86-991-3674558

Academic Editor: Stevin Snellius Pramana
Received: 23 February 2017; Accepted: 6 April 2017; Published: 12 April 2017

Abstract: A new family of quaternary diamond-like semiconductors (DLSs), Li$_2$HgMS$_4$ (M = Si, Ge, Sn), were successfully discovered for the first time. All of them are isostructural and crystallize in the polar space group ($Pmn2_1$). Seen from their structures, they exhibit a three-dimensional (3D) framework structure that is composed of countless 2D honeycomb layers stacked along the c axis. An interesting feature, specifically, that the LiS$_4$ tetrahedra connect with each other to build a 2D layer in the ac plane, is also observed. Experimental investigations show that their nonlinear optical responses are about 0.8 for Li$_2$HgSiS$_4$, 3.0 for Li$_2$HgGeS$_4$, and 4.0 for Li$_2$HgSnS$_4$ times that of benchmark AgGaS$_2$ at the 55–88 μm particle size, respectively. In addition, Li$_2$HgSiS$_4$ and Li$_2$HgGeS$_4$ also have great laser-damage thresholds that are about 3.0 and 2.3 times that of powdered AgGaS$_2$, respectively. The above results indicate that title compounds can be expected as promising IR NLO candidates.

Keywords: nonlinear optical materials; crystal structure; good NLO responses

---

1. Introduction

Solid-state lasers have shown a wide range of applications in the fields of military, industry, medical treatment and information communications [1,2]. However, traditional laser sources, such as Ti:Al$_2$O$_3$ and Nd:YAG lasers, mainly cover the wavelengths range from visible to near infrared, not including the important ultraviolet (UV < 400 nm) and middle-far infrared (MFIR, 3–20 μm) region [3,4]. To extend the laser wavelength ranges, frequency-conversion technology on nonlinear optical (NLO) materials was invented and has been further developed for decades [5]. Recently, many promising NLO materials have been discovered and have basically solved the demand of UV region [6–34]. However, for the IR region, outstanding IR NLO materials were rarely discovered and only several ternary diamond-like semiconductors (DLSs), such as AgGaS$_2$, AgGaSe$_2$ and ZnGeP$_2$, have been commercially used [35–37]. Although they have high second harmonic generation (SHG) coefficients and wide IR transmission regions, some of the self-defects including the low laser-damage thresholds (LDTs) or strong two-photon absorption (TPA) still seriously hinder their practical application. Researchers have done a lot of work to explore new NLO materials for the IR application, and the combination of two or more different building units into crystal structures can be viewed as a feasible way to obtain new NLO compounds. Up to now, many reports indicate that cations with second order Jahn–Teller distortions, lone electron pairs or d^{10} configuration can contribute to good SHG response with the cooperative effects of typical tetrahedral units MQ_4 (M = Ga, In, Si, Ge, Sn; Q = S, Se) [38–73]. Note that diamond-like semiconductors (DLSs) with inherently noncentrosymmetrical (NCS) structures, with all the tetrahedral units in crystal structures oriented in the same direction, have been further investigated on the ternary and quaternary systems [74–80]. Among them, the d^{10} cations (Zn^{2+}, Cd^{2+}) containing quaternary DLSs with outstanding performances, such as Li_2CdGeS_4, $Li_2ZnGeSe_4$, and $Li_2ZnSnSe_4$ [74,78], were proven as promising IR NLO candidates and belong to the general formula I_2–II–IV–VI_4, where I are the monovalent elements, II are the divalent elements, IV are the group 14 elements, and VI are the chalcogen elements. Previous reports indicate that other types of Hg-containing metal chalcogenides have shown good NLO performances in the IR field, such as $HgGaS_4$ [81], $BaHgQ_2$ (Q = S, Se), [82,83] $A_2Hg_3M_2S_8$ (A = Na, K; M = Si, Ge, Sn) [61,68], and $Li_4HgGe_2S_7$ [84]. However, up to now, the Hg-containing DLSs have been rarely investigated in the IR frequency conversion region. Thus, it is meaningful to explore new Hg-containing DLSs and study their important NLO properties. From this background, we have chosen the Li–Hg–M–S (M = Si, Ge, Sn) as the research system and successfully prepared three new IR NLO materials, Li_2HgMS_4 (M = Si, Ge, Sn). They are isostructural and crystallize in the $Pmn2_1$ polar space group. Overall properties investigation shows that they can be expected to be promising IR NLO candidates owing to their large NLO coefficients, high LDTs, wide IR transparent regions, and good chemical stability.

2. Results and Discussion

2.1. Crystal Structure

The title compounds are isostructural and crystallize in the NCS polar space group $Pmn2_1$. In order to ensure the reasonability of crystal structures of title compounds, the bond valence [85,86] and the global instability index (GII) [87–89] were also systemically studied (Table 1). Calculated results (Li, 1.085–1.125; Hg, 2.090–2.133; Si/Ge/Sn, 4.030–4.152; S, 2.055–2.229) indicate that all atoms are in reasonable oxidation states. In addition, GII can be derived from the bond valence concepts, which represent the tension of lattice parameters and are always used to evaluate the rationality of structure. When the value of GII is less than 0.05 vu (valence unit), the tension of structure is not proper, whereas when the value of GII is larger than 0.2 vu, its structure is not stable. Thus, the value of GII in a reliable structure should be limited at 0.05–0.2 in general. As for the title compounds, calculated GII values are in the range of 0.10–0.14 vu, which illustrates that the crystal structures of all compounds are reasonable.

Table 1. Bond Valence Sum (vu) and Global Instability Index (GII) of title compounds.

Compounds	Li^+	Hg^{2+}	$Si/Ge/Sn^{4+}$	S^{2-}	GII
Li_2HgSiS_4	1.125	2.092	4.030	2.069–2.149	0.10
Li_2HgGeS_4	1.118	2.090	4.152	2.055–2.229	0.14
Li_2HgSnS_4	1.085	2.133	4.138	2.061–2.163	0.13

Herein, Li_2HgGeS_4 is chosen as the representative for the structural discussion. In its structure, each cation is linked to four S atoms, forming the typical LiS_4, HgS_4, and GeS_4 tetrahedra. These units connect with each other to make up a two-dimensional (2D) honeycomb layer structure, which is located at the *ab* plane (Figure 1b). Then, the layers are further stacked along the *c* axis to form a three-dimensional (3D) framework structure (Figure 1a). In addition, an interesting feature is that the LiS_4 tetrahedra connect with each other to build a 2D layer in the *ac* plane (Figure 1c). The whole structure is composed of tetrahedral ligands that align along the *c* axis. Note that the discovered quaternary DLSs in the I_2–II–IV–VI_4 systems normally crystallize in the one of following space groups: I-$42m$ (Cu_2CdSnS_4) [76], I-4 (Cu_2ZnSnS_4) [90], $Pmn2_1$ (Li_2CdGeS_4) [74], $Pna2_1$ (Li_2MnGeS_4) [75], and Pn (Li_2CoSnS_4) [75], to our best knowledge, which represent the stannite, kesterite, wurtz-stannite, and wurtz-kesterite structural features.

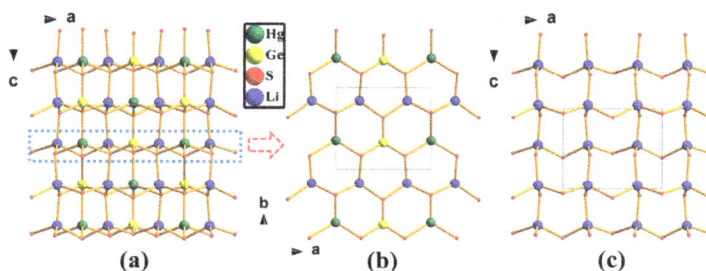

Figure 1. (**a**) View of the crystal structure of Li$_2$HgGeS$_4$ along the *b* axis; (**b**) A honeycomb-like layer composed of the LiS$_4$, HgS$_4$, and GeS$_4$ units located at the *ab* plane; (**c**) A layer composed of the LiS$_4$ units located at the *ac* plane.

2.2. Optical Properties

As for an IR NLO crystal, its optical parameters, such as optical bandgap, IR absorption edge, NLO response, and LDT, are necessary to be determined for the assessment of application prospect. The detailed mechanism for laser damage in a given material is not fully clear yet, however it has been normally accepted that strong optical absorption of the materials will cause thermal and electronic effects and finally lead to laser damage. Note that the optical breakdown can be attributed to the effect of electron avalanche that has a close relationship with optical bandgap in a given material [75]. Figure 2 shows that the optical bandgaps are 2.68 for Li$_2$HgSiS$_4$, 2.46 for Li$_2$HgGeS$_4$, and 2.32 eV for Li$_2$HgSnS$_4$, respectively. All of them are much larger than those of commercial AgGaSe$_2$ (1.80 eV) [86] and ZnGeP$_2$ (1.75 eV) [37], which may be conducive to improve the laser damage resistance of the title compounds compared with the commercially available IR NLO materials. Recently, the assessment of the LDTs on powder samples has been developed as a feasible and semi-quantitative method [52,55]. Thus, in this work, based on a pulse laser (1.06 μm, 10 Hz and 10 ns), the LDTs of the title compounds were measured with AgGaS$_2$ as the reference and corresponding results are shown in Table 2. From this table, it can be found that the title compounds have great LDTs, such as 91.6 for Li$_2$HgSiS$_4$, 70.6 for Li$_2$HgGeS$_4$, and 30.5 MW/cm^2 for Li$_2$HgSnS$_4$, and are about 3.0, 2.3, and ~1 times that of powdered AgGaS$_2$ (29.6 MW/cm^2), respectively. Moreover, Li$_2$HgSiS$_4$ and Li$_2$HgGeS$_4$ are comparable to those of PbGa$_2$GeSe$_6$ (3.7 × AgGaS$_2$) [51], Na$_2$In$_2$GeS$_6$ (4.0 × AgGaS$_2$) [56], Na$_2$Hg$_3$Ge$_2$S$_8$ (3 × AgGaS$_2$) [61], and SnGa$_4$Se$_7$ (4.6 × AgGaS$_2$) [52]. Therefore, Li$_2$HgSiS$_4$ and Li$_2$HgGeS$_4$ can be expected to have application with high-power lasers, compared with the commercial IR NLO materials.

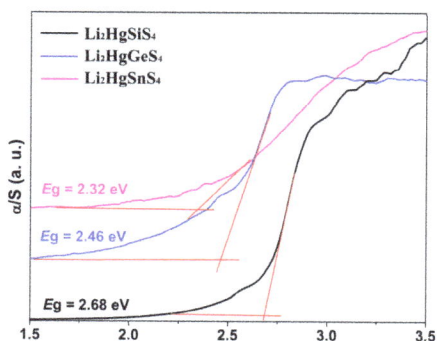

Figure 2. Experimental bandgaps of the title compounds.

Table 2. LDTs of the title compounds and AgGaS$_2$ (as the reference).

Compounds	Damage Energy (mJ)	Spot Diameter (mm)	LDT (MW/cm^2)
AgGaS$_2$	0.33	0.375	29.6
Li$_2$HgSiS$_4$	1.02	0.375	91.6
Li$_2$HgGeS$_4$	0.78	0.375	70.2
Li$_2$HgSnS$_4$	0.34	0.375	30.5

In addition, Raman spectra (Figure 3) are also measured to determine the IR absorption edges for the title compounds. The results show that all of them exhibit the wide IR transmission regions, such as 2.5–19 μm (530 cm^{-1}) for Li$_2$HgSiS$_4$, 2.5–22 μm (450 cm^{-1}) for Li$_2$HgGeS$_4$, and 2.5–23.5 μm (425 cm^{-1}) for Li$_2$HgSnS$_4$, which cover the two important atmospheric windows (3–5 and 8–12 μm) that can be used in telecommunications, laser guidance, and explosives detection. Note that the IR absorption edges gradually get longer from the Si to Sn compounds, which are consistent with the IR data for other related mental chalcogenides [61]. Although the measured IR absorption data on powder samples have some deviations with the results on single-crystals, they can give the preliminary assessment for the transmission region of IR materials. Overall view on Raman spectra shows similar patterns for the title compounds, and a shift to lower absorption energies from the Si to Sn compounds that are severely affected by the tetravalent (M^{IV}) metals. The absorption peaks above approximately 300 cm^{-1}, including Li$_2$HgSiS$_4$ (520, 396 cm^{-1}), Li$_2$HgGeS$_4$ (430, 387, 360, 327 cm^{-1}), and Li$_2$HgSnS$_4$ (402, 346 cm^{-1}), can be assigned to the characteristic absorptions of the Si–S, Ge–S, and Sn–S modes, respectively. Moreover, several peaks located between 200 and 300 cm^{-1}, such as Li$_2$HgSiS$_4$ (285, 258 cm^{-1}), Li$_2$HgGeS$_4$ (257 cm^{-1}), and Li$_2$HgSnS$_4$ (258 cm^{-1}), are attributed to the Hg–S bonding interactions. In addition, the absorptions below 200 cm^{-1} are primarily corresponding to the Li–S vibrations for the title compounds.

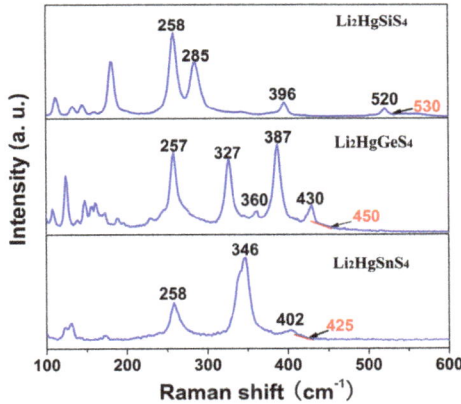

Figure 3. Raman spectra of the title compounds.

Second harmonic generation (SHG) responses for the title compounds were investigated on powder samples and the results are shown in Figure 4. From this figure, it can be found that the SHG intensities of the title compounds are not enhanced gradually with the increase of particle sizes, which indicates the nonphase matching behaviour for these compounds. In addition, their SHG responses are about 0.8 for Li$_2$HgSiS$_4$, 3.0 for Li$_2$HgGeS$_4$, and 4.0 for Li$_2$HgSnS$_4$ times that of benchmark AgGaS$_2$ at the 55–88 μm particle size, respectively, which shows that the title compounds may have great NLO potential in the IR region as promising frequency-conversion candidates.

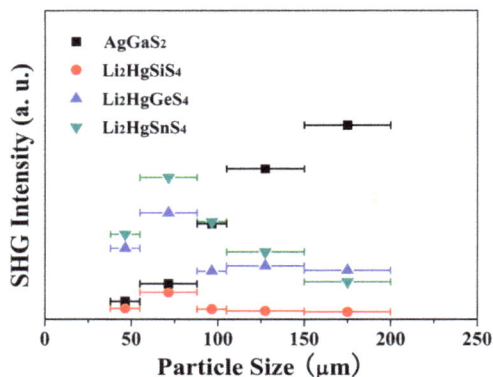

Figure 4. Second-harmonic generation (SHG) intensity versus particle size for the title compounds and $AgGaS_2$.

3. Materials and Methods

3.1. Synthesis

All the starting materials were used as purchased without further refinement. In the preparation process, a graphite crucible was added into the vacuum sealed silica tube to avoid the reaction between metal Li and the silica tube at a high temperature.

3.1.1. Li_2HgSiS_4 and Li_2HgSnS_4

Target compounds were prepared with a mixture with the ratio of Li:HgS:(Si or Sn):S = 2:1:1:3, respectively. The temperature process was set as follows: first, it was heated to 700 °C in two days, and kept at this temperature about four days, then slowly down to 300 °C within four days, and finally quickly cooled to room temperature by turning off the furnace. Obtained products were washed by the *N,N*-dimethylformamide (DMF) solvent to remove the other byproducts. Yellow crystals for Li_2HgSiS_4 and orange-red crystals for Li_2HgSnS_4 appeared, and both of them remained stable in air over half a year. In addition, the yield of Li_2HgSiS_4 was about 80%.

3.1.2. Li_2HgGeS_4

Initially, we attempted to prepare Li_2HgGeS_4 with the ratio of Li:HgS:Ge:S = 2:1:1:3 at the reaction temperature of 700 °C. After the single crystal X-ray diffraction measurement, $Li_4HgGe_2S_7$ (main product, yellow) [79] and Li_2HgGeS_4 (a small amount, reddish) were interestingly obtained. In addition, we had further adjusted the ratio of reactants and interestingly found that the pure-phase of Li_2HgGeS_4 could be obtained while the ratio of Li:HgS is greater than 2:1. Moreover, the Li_2HgGeS_4 crystals were repeatedly washed with DMF solvent and they also remained stable in air.

3.2. Structure Determination

Selected single-crystals were used for data collections with a Bruker SMART APEX II 4K CCD diffractometer (Bruker Corporation, Madison, WI, USA) using Mo Kα radiation (λ = 0.71073 Å) at room temperature. Multi-scan method was used for absorption correction [91]. All the crystal structures were solved by the direct method and refined using the SHELXTL program package [92]. As for the structural refinement of Li_2HgSiS_4, the initial refinement result gave the formula Li_2HgSiS_4, but the site of the Li atom showed abnormal anisotropy parameter (almost zero). Thus, we attempted to set the Li1 and Hg2 atoms to occupy the same site with the ratio of 0.97:0.03 (Li1:Hg2) by random refinement. In view of the low occupancy (0.025) of Hg2 atom, we consider using the "ISOR" order to treat the Li1 atom as isotropic instead of a positional disorder (Li1:Hg2). Moreover, the subsequent analysis of the element contents in the title compounds with energy dispersive X-ray (EDX) equipped Hitachi S-4800 SEM (Tokyo, Japan) showed the approximate molar ratio of 1:1:4 for Hg, Si/Ge/Sn, and S (Li is undetectable in EDX). The final structures were carefully checked with PLATON software (Glasgow, UK) and no other symmetries were found [93]. Table 3 shows the crystal data and structure refinement of the title compounds.

Table 3. Crystal data and structure refinement for the title compounds.

Empirical Formula	Li_2HgSiS_4	Li_2HgGeS_4	Li_2HgSnS_4
fw	370.80	415.30	461.40
crystal system	*orthorhombic*	*orthorhombic*	*orthorhombic*
space group	*Pmn2$_1$*	*Pmn2$_1$*	*Pmn2$_1$*
a (Å)	7.592 (2)	7.709 (9)	7.9400 (17)
b (Å)	6.7625 (19)	6.812 (8)	6.9310 (15)
c (Å)	6.3295 (18)	6.384 (7)	6.5122 (14)
Z, V (Å^3)	2, 324.96 (16)	2, 335.3 (7)	2, 358.38 (13)
D_c (g/cm^3)	3.790	4.114	4.276
μ (mm^{-1})	25.014	28.463	25.918
GOF on F^2	1.022	1.161	0.985
R_1, wR_2 (I > 2σ(I)) ^{a}	0.0217, 0.0443	0.0422, 0.0994	0.0318, 0.0633
R_1, wR_2 (all data)	0.0229, 0.0445	0.0438, 0.0999	0.0423, 0.0682
absolute structure parameter	0.003 (11)	0.04 (3)	−0.019 (19)
largest diff. peak and hole (e Å^{-3})	1.318, −1.170	5.723, −1.070	0.959, −1.797

^{a} $R_1 = F_o - F_c/F_o$ and $wR_2 = [w (F_o{}^2 - F_c{}^2)^2/wF_o{}^4]^{1/2}$ for $F_o{}^2 > 2\sigma (F_o{}^2)$.

3.3. Powder XRD Measurement

A Bruker D2 X-ray diffractometer (Madison, USA) with Cu Kα radiation (λ = 1.5418 Å) was used to measure the powder X-ray diffraction (XRD) patterns of title compounds at room temperature. The measured range is 10–70° with a step size of 0.02°. Compared with the calculated and experiment results, it can be concluded that they are basically consistent with each other, except for Li_2HgSiS_4 with a small number of the Hg_4SiS_4 impurities (Figure 5).

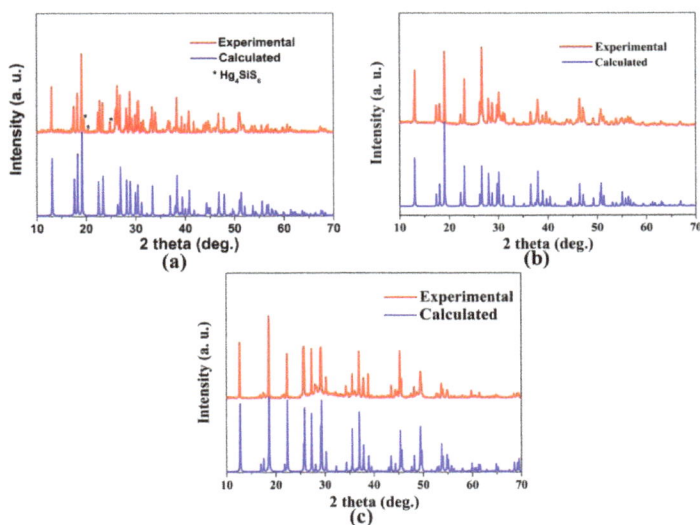

Figure 5. Powder XRD patterns of Li_2HgSiS_4 (**a**), Li_2HgGeS_4 (**b**), Li_2HgSnS_4 (**c**).

3.4. UV–Vis–NIR Diffuse-Reflectance Spectroscopy

Diffuse-reflectance spectra were measured by a Shimadzu SolidSpec-3700DUV spectrophotometer (Shimadzu Corporation, Beijing, China) in the wavelength range of 190–2600 nm at room temperature. The absorption spectra were converted from the reflection spectra via the Kubelka–Munk function.

3.5. Raman Spectroscopy

Hand-picked crystals were first put on an object slide, and then a LABRAM HR Evolution spectrometer equipped with a CCD detector (HORIBA Scientific, Beijing, China) by a 532-nm laser was used to record the Raman spectra. The integration time was set to be 10 s.

3.6. Second-Harmonic Generation Measurement

By the Kurtz and Perry method, powder SHG responses of the title compounds were investigated by a Q-switch laser (2.09 μm, 3 Hz, 50 ns) with ground micro-crystals on different particle sizes. $AgGaS_2$ single-crystal was also ground and sieved into the same size range as the reference. SHG signals were detected by a digital oscilloscope.

3.7. LDT Measurement

Ground micro-crystals samples (55–88 μm) were used to evaluate the LDTs of the title compounds under a pulsed YAG laser (1.06 μm, 10 ns, 10 Hz). Similar sizes of the $AgGaS_2$ crystal were chosen as the reference. By adjusting the laser output energy, colour change of the test sample was carefully observed by an optical microscope to determine the LDTs.

4. Conclusions

A new family of new DLSs, Li_2HgMS_4 (M = Si, Ge, Sn), were successfully synthesized by the solid-state method in vacuum-sealed silica tubes. They are isostructural and crystallize in the orthorhombic $Pmn2_1$ space group. Seen from their structures, they have the similar 3D framework and 2D honeycomb-like layer structures with the interconnection of three types of tetrahedral units (LiS_4, HgS_4, and MS_4). Corresponding optical properties for the title compounds are systemically studied and the results show that they have great potential as promising IR NLO candidates.

Supplementary Materials: The following are available online at www.mdpi.com/2073-4352/7/4/107/s1. Cifs for title compounds.

Acknowledgments: This work was supported by the Western Light Foundation of CAS (Grant No. XBBS201318), the National Natural Science Foundation of China (Grant Nos. 51402352, 51425206, 91622107), Fund of Key Laboratory of Optoelectronic Materials Chemistry and Physics, Chinese Academy of Sciences (2008DP173016).

Author Contributions: Kui Wu conceived and designed this study, prepared the crystals and wrote the manuscript. Shilie Pan conceived and coordinated the project.

Conflicts of Interest: The authors declare no conflict of interest.

References

1. Keller, U. Recent developments in compact ultrafast lasers. *Nature* **2003**, *424*, 831–838. [CrossRef] [PubMed]
2. Byer, R.L. Diode laser-pumped solid-state lasers. *Science* **1988**, *239*, 742–748. [CrossRef] [PubMed]
3. Duarte, F.J. *Tunable Laser Applications*, 2nd ed.; CRC Press: Boca Raton, FL, USA, 2008; Chapters 2, 9 and 12.
4. Demtröder, W. *Laser Spectroscopy*, 3rd ed.; Springer: Berlin, Germany, 2009.
5. Nikogosyan, D.N. *Nonlinear Optical Crystals: A Complete Survey*, 1st ed.; Springer: New York, NY, USA, 2005.
6. Chen, C.T.; Wu, Y.C.; Jiang, A.D.; Wu, B.C.; You, G.M.; Li, R.K.; Lin, S.J. New Nonlinear-Optical Crystal: LiB_3O_5. *J. Opt. Soc. Am. B* **1989**, *6*, 616–621. [CrossRef]
7. Chen, C.T.; Wu, B.C.; Jiang, A.D.; You, G.M. A New-Type Ultraviolet SHG Crystal-β-BaB_2O_4. *Sci. Sin. Ser. B* **1985**, *28*, 235–243.

8. Mei, L.; Wang, Y.; Chen, C.T.; Wu, B.C. Nonlinear Optical Materials Based on MBe$_2$BO$_3$F$_2$ (M = Na, K). *J. Appl. Phys.* **1993**, *74*, 7014–7016. [CrossRef]

9. Wang, G.L.; Zhang, C.Q.; Chen, C.T.; Yao, A.Y.; Zhang, J.; Xu, Z.Y.; Wang, J.Y. High-Efficiency 266-nm Output of a KBe$_2$BO$_3$F$_2$ Crystal. *Appl. Opt.* **2003**, *42*, 4331–4334. [CrossRef] [PubMed]

10. Becker, P. Borate Materials in Nonlinear Optics. *Adv. Mater.* **1998**, *10*, 979–992. [CrossRef]

11. Sun, C.F.; Hu, C.L.; Xu, X.; Ling, J.B.; Hu, T.; Kong, F.; Long, X.F.; Mao, J.G. BaNbO(IO$_3$)$_5$: A New Polar Material with A Very Large SHG Response. *J. Am. Chem. Soc.* **2009**, *131*, 9486–9487. [CrossRef] [PubMed]

12. Hu, C.L.; Mao, J.G. Recent Advances on Second-Order NLO Materials Based on Metal Iodates. *Coord. Chem. Rev.* **2015**, *288*, 1–17. [CrossRef]

13. Song, J.L.; Hu, C.L.; Xu, X.; Kong, F.; Mao, J.G. A Facile Synthetic Route to a New SHG Material with Two Types of Parallel π-Conjugated Planar Triangular Units. *Angew. Chem. Int. Ed.* **2015**, *54*, 3679–3682. [CrossRef] [PubMed]

14. Zou, G.H.; Ye, N.; Huang, L.; Lin, X.S. Alkaline-Alkaline Earth Fluoride Carbonate Crystals ABCO$_3$F (A = K, Rb, Cs; B = Ca, Sr, Ba) as Nonlinear Optical Materials. *J. Am. Chem. Soc.* **2011**, *133*, 20001–20007. [CrossRef] [PubMed]

15. Yang, G.S.; Peng, G.; Ye, N.; Wang, J.Y.; Luo, M.; Yan, T.; Zhou, Y.Q. Structural Modulation of Anionic Group Architectures by Cations to Optimize SHG Effects: A Facile Route to New NLO Materials in the ATCO$_3$F (A = K, Rb; T = Zn, Cd) Series. *Chem. Mater.* **2015**, *27*, 7520–7530. [CrossRef]

16. Zou, G.H.; Huang, L.; Ye, N.; Lin, C.S.; Cheng, W.D.; Huang, H. CsPbCO$_3$F: A Strong Second-Harmonic Generation Material Derived from Enhancement via p–π Interaction. *J. Am. Chem. Soc.* **2013**, *135*, 18560–18566. [CrossRef] [PubMed]

17. Yu, H.W.; Zhang, W.G.; Young, J.; Rondinelli, J.M.; Halasyamani, P.S. Bidenticity-Enhanced Second Harmonic Generation from Pb Chelation in Pb$_3$Mg$_3$TeP$_2$O$_{14}$. *J. Am. Chem. Soc.* **2016**, *138*, 88–91. [CrossRef] [PubMed]

18. Yu, H.W.; Zhang, W.G.; Young, J.; Rondinelli, J.M.; Halasyamani, P.S. Design and Synthesis of the Beryllium-Free Deep-Ultraviolet Nonlinear Optical Material Ba$_3$(ZnB$_5$O$_{10}$)PO$_4$. *Adv. Mater.* **2015**, *27*, 7380–7385. [CrossRef] [PubMed]

19. Kim, H.G.; Tran, T.T.; Choi, W.; You, T.S.; Halasyamani, P.S.; Ok, K.M. Two New Non-centrosymmetric *n* = 3 Layered Dion–Jacobson Perovskites: Polar RbBi$_2$Ti$_2$NbO$_{10}$ and Nonpolar CsBi$_2$Ti$_2$TaO$_{10}$. *Chem. Mater.* **2016**, *28*, 2424–2432. [CrossRef]

20. Zou, G.H.; Nam, G.; Kim, H.G.; Jo, H.; You, T.S.; Ok, K.M. ACdCO$_3$F (A = K and Rb): New Noncentrosymmetric Materials with Remarkably Strong Second-Harmonic Generation (SHG) Responses Enhanced via π-Interaction. *RSC Adv.* **2015**, *5*, 84754–84761. [CrossRef]

21. Cheng, L.; Wei, Q.; Wu, H.Q.; Zhou, L.J.; Yang, G.Y. Ba$_3$M$_2$[B$_3$O$_6$(OH)]$_2$[B$_4$O$_7$(OH)$_2$] (M = Al, Ga): Two Novel UV Nonlinear Optical Metal Borates Containing Two Types of Oxoboron Clusters. *Chem. Eur. J.* **2013**, *19*, 17662–17667. [CrossRef] [PubMed]

22. Huang, H.W.; Liu, L.J.; Jin, S.F.; Yao, W.J.; Zhang, Y.H.; Chen, C.T. Deep-Ultraviolet Nonlinear Optical Materials: Na$_2$Be$_4$B$_4$O$_{11}$ and LiNa$_5$Be$_{12}$B$_{12}$O$_{33}$. *J. Am. Chem. Soc.* **2013**, *135*, 18319–18322. [CrossRef] [PubMed]

23. Li, F.; Hou, X.L.; Pan, S.L.; Wang, X.A. Growth, structure, and optical properties of a congruent melting oxyborate, Bi$_2$ZnOB$_2$O$_6$. *Chem. Mater.* **2009**, *21*, 2846–2850. [CrossRef]

24. Wu, H.P.; Pan, S.L.; Poeppelmeier, K.R.; Li, H.Y.; Jia, D.Z.; Chen, Z.H.; Fan, X.Y.; Yang, Y.; Rondinelli, J.M.; Luo, H. K$_3$B$_6$O$_{10}$Cl: A New Structure Analogous to Perovskite with A Large Second Harmonic Generation Response and Deep UV Absorption Edge. *J. Am. Chem. Soc.* **2011**, *133*, 7786–7790. [CrossRef] [PubMed]

25. Yu, H.W.; Wu, H.P.; Pan, S.L.; Yang, Z.H.; Hou, X.L.; Su, X.; Jing, Q.; Poeppelmeier, K.R.; Rondinelli, J.M. Cs$_3$Zn$_6$B$_9$O$_{21}$: A Chemically Benign Member of the KBBF Family Exhibiting the Largest Second Harmonic Generation Response. *J. Am. Chem. Soc.* **2014**, *136*, 1264–1267. [CrossRef] [PubMed]

26. Li, L.; Wang, Y.; Lei, B.H.; Han, S.J.; Yang, Z.H.; Poeppelmeier, K.R.; Pan, S.L. A New Deep-Ultraviolet Transparent Orthophosphate LiCs$_2$PO$_4$ with Large Second Harmonic Generation Response. *J. Am. Chem. Soc.* **2016**, *138*, 9101–9104. [CrossRef] [PubMed]

27. Wang, Y.; Pan, S.L. Recent Development of Metal Borate Halides: Crystal Chemistry and Application in Second-order NLO Materials. *Coord. Chem. Rev.* **2016**, *323*, 15–35. [CrossRef]

28. Dong, X.Y.; Jing, Q.; Shi, Y.J.; Yang, Z.H.; Pan, S.L.; Poeppelmeier, K.R.; Young, J.S.; Rondinelli, J.M. $Pb_2Ba_3(BO_3)_3Cl$: A Material with Large SHG Enhancement Activated by Pb-Chelated BO_3 Groups. *J. Am. Chem. Soc.* **2015**, *137*, 9417–9422. [CrossRef] [PubMed]

29. Wu, H.P.; Yu, H.W.; Pan, S.L.; Huang, Z.J.; Yang, Z.H.; Su, X.; Poeppelmeier, K.R. $Cs_2B_4SiO_9$, A Deep-ultraviolet Nonlinear Optical Crystal. *Angew. Chem. Int. Ed.* **2013**, *52*, 3406–3410. [CrossRef] [PubMed]

30. Wu, H.P.; Yu, H.W.; Yang, Z.H.; Hou, X.L.; Su, X.; Pan, S.L.; Poeppelmeier, K.R.; Rondinelli, J.M. Designing a Deep-Ultraviolet Nonlinear Optical Material with a Large Second Harmonic Generation Response. *J. Am. Chem. Soc.* **2013**, *135*, 4215–4218. [CrossRef] [PubMed]

31. Zhao, S.; Gong, P.; Bai, L.; Xu, X.; Zhang, S.; Sun, Z.; Lin, Z.; Hong, M.; Chen, C.; Luo, J. Beryllium-free $Li_4Sr(BO_3)_2$ for Deep-ultraviolet Nonlinear Optical Applications. *Nat. Commun.* **2014**, *5*. [CrossRef] [PubMed]

32. Zhao, S.G.; Gong, P.F.; Luo, S.Y.; Bai, L.; Lin, Z.S.; Tang, Y.Y.; Zhou, Y.L.; Hong, M.C.; Luo, J.H. Tailored Synthesis of a Nonlinear Optical Phosphate with a Short Absorption Edge. *Angew. Chem. Int. Ed.* **2015**, *54*, 4217–4221. [CrossRef] [PubMed]

33. Zhao, S.G.; Gong, P.F.; Luo, S.Y.; Liu, S.J.; Li, L.N.; Asghar, M.A.; Khan, T.; Hong, M.C.; Lin, Z.S.; Luo, J.H. Beryllium-Free $Rb_3Al_3B_3O_{10}F$ with Reinforced Inter layer Bonding as a Deep-Ultraviolet Nonlinear Optical Crystal. *J. Am. Chem. Soc.* **2015**, *137*, 2207–2210. [CrossRef] [PubMed]

34. Jiang, X.X.; Luo, S.Y.; Kang, L.; Gong, P.F.; Huang, H.W.; Wang, S.C.; Lin, Z.S.; Chen, C.T. First-Principles Evaluation of the Alkali and/or Alkaline Earth Beryllium Borates in Deep Ultraviolet Nonlinear Optical Applications. *ACS Photonics* **2015**, *2*, 1183–1191. [CrossRef]

35. Okorogu, A.O.; Mirov, S.B.; Lee, W.; Crouthamel, D.I.; Jenkins, N.; Dergachev, A.Y.; Vodopyanov, K.L.; Badikov, V.V. Tunable Middle Infrared Downconversion in GaSe and $AgGaS_2$. *Opt. Commun.* **1998**, *155*, 307–312. [CrossRef]

36. Boyd, G.D.; Storz, F.G.; McFee, J.H.; Kasper, H.M. Linear and Nonlinear Optical Properties of Some Ternary Selenides. *IEEE J. Quantum Electron.* **1972**, *8*, 900–908. [CrossRef]

37. Boyd, G.D.; Buehler, E.; Storz, F.G. Linear and Nonlinear Optical Properties of $ZnGeP_2$ and CdSe. *Appl. Phys. Lett.* **1971**, *18*, 301–304. [CrossRef]

38. Chung, I.; Kanatzidis, M.G. Metal chalcogenides: A rich source of nonlinear optical materials. *Chem. Mater.* **2014**, *26*, 849–869. [CrossRef]

39. Jiang, X.M.; Guo, S.P.; Zeng, H.Y.; Zhang, M.J.; Guo, G.C. Large Crystal Growth and New Crystal Exploration of Mid-Infrared Second-Order Nonlinear Optical Materials. *Struct. Bond.* **2012**, *145*, 1–43.

40. Liang, F.; Kang, L.; Lin, Z.S.; Wu, Y.C.; Chen, C.T. Analysis and Prediction of Mid-IR Nonlinear Optical Metal Sulfides with Diamond-like Structures. *Coord. Chem. Rev.* **2017**, *333*, 57–70. [CrossRef]

41. Guo, S.P.; Chi, Y.; Guo, G.C. Recent Achievements on Middle and Far-infrared Second-order Nonlinear Optical Materials. *Coord. Chem. Rev.* **2017**, *335*, 44–57. [CrossRef]

42. Bera, T.K.; Jang, J.I.; Ketterson, J.B.; Kanatzidis, M.G. Strong Second Harmonic Generation from the Tantalum Thioarsenates $A_3Ta_2AsS_{11}$ (A = K and Rb). *J. Am. Chem. Soc.* **2009**, *131*, 75–77. [CrossRef] [PubMed]

43. Zhang, W.; Li, F.; Kim, S.H.; Halasyamani, P.S. Top-Seeded Solution Crystal Growth and Functional Properties of a Polar Material $Na_2TeW_2O_9$. *Cryst. Growth Des.* **2010**, *10*, 4091–4095. [CrossRef]

44. Lin, H.; Zhou, L.J.; Chen, L. Sulfides with Strong Nonlinear Optical Activity and Thermochromism: $ACd_4Ga_5S_{12}$ (A = K, Rb, Cs). *Chem. Mater.* **2012**, *24*, 3406–3414. [CrossRef]

45. Chen, M.C.; Wu, L.M.; Lin, H.; Zhou, L.J.; Chen, L. Disconnection Enhances the Second Harmonic Generation Response: Synthesis and Characterization of $Ba_{23}Ga_8Sb_2S_{38}$. *J. Am. Chem. Soc.* **2012**, *134*, 6058–6060. [CrossRef] [PubMed]

46. Yu, P.; Zhou, L.J.; Chen, L. Noncentrosymmetric Inorganic Open-Framework Chalcohalides with Strong Middle IR SHG and Red emission: $Ba_3AGa_5Se_{10}Cl_2$ (A = Cs, Rb, K). *J. Am. Chem. Soc.* **2012**, *134*, 2227–2235. [CrossRef] [PubMed]

47. Chen, M.C.; Li, L.H.; Chen, Y.B.; Chen, L. In-Phase Alignments of Asymmetric Building Units in Ln_4GaSbS_9 (Ln = Pr, Nd, Sm, Gd–Ho) and Their Strong Nonlinear Optical Responses in Middle IR. *J. Am. Chem. Soc.* **2011**, *133*, 4617–4624. [CrossRef] [PubMed]

48. Chen, Y.K.; Chen, M.C.; Zhou, L.J.; Chen, L.; Wu, L.M. Syntheses, Structures, and Nonlinear Optical Properties of Quaternary Chalcogenides: $Pb_4Ga_4GeQ_{12}$ (Q = S, Se). *Inorg. Chem.* **2013**, *52*, 8334–8341. [CrossRef] [PubMed]

49. Yao, J.Y.; Mei, D.J.; Bai, L.; Lin, Z.S.; Yin, W.L.; Fu, P.Z.; Wu, Y.C. $BaGa_4Se_7$: A New Congruent-Melting IR Nonlinear Optical Material. *Inorg. Chem.* **2010**, *49*, 9212–9216. [CrossRef] [PubMed]

50. Lin, X.S.; Zhang, G.; Ye, N. Growth and Characterization of $BaGa_4S_7$: A New Crystal for Mid-IR Nonlinear Optics. *Cryst. Growth Des.* **2009**, *9*, 1186–1189. [CrossRef]

51. Luo, Z.Z.; Lin, C.S.; Cui, H.H.; Zhang, W.L.; Zhang, H.; Chen, H.; He, Z.Z.; Cheng, W.D. $PbGa_2MSe_6$ (M = Si, Ge): Two Exceptional Infrared Nonlinear Optical Crystals. *Chem. Mater.* **2015**, *27*, 914–922. [CrossRef]

52. Luo, Z.Z.; Lin, C.S.; Cui, H.H.; Zhang, W.L.; Zhang, H.; He, Z.Z.; Cheng, W.D. SHG Materials $SnGa_4Q_7$ (Q = S, Se) Appearing with Large Conversion Efficiencies, High Damage Thresholds, and Wide Transparencies in the Mid-Infrared Region. *Chem. Mater.* **2014**, *26*, 2743–2749. [CrossRef]

53. Geng, L.; Cheng, W.D.; Lin, C.S.; Zhang, W.L.; Zhang, H.; He, Z.Z. Syntheses and Characterization of New Mid-Infrared Transparency Compounds: Centric Ba_2BiGaS_5 and Acentric Ba_2BiInS_5. *Inorg. Chem.* **2011**, *50*, 5679–5686. [CrossRef] [PubMed]

54. Luo, Z.Z.; Lin, C.S.; Zhang, W.L.; Zhang, H.; He, Z.Z.; Cheng, W.D. $Ba_8Sn_4S_{15}$: A Strong Second Harmonic Generation Sulfide with Zero-Dimensional Crystal Structure. *Chem. Mater.* **2013**, *26*, 1093–1099. [CrossRef]

55. Liu, B.W.; Zeng, H.Y.; Zhang, M.J.; Fan, Y.H.; Guo, G.C.; Huang, J.S.; Dong, Z.C. Syntheses, Structures, and Nonlinear-optical Properties of Metal Sulfides $Ba_2Ga_8MS_{16}$ (M = Si, Ge). *Inorg. Chem.* **2014**, *54*, 976–981. [CrossRef] [PubMed]

56. Li, S.F.; Liu, B.W.; Zhang, M.J.; Fan, Y.H.; Zeng, H.Y.; Guo, G.C. Syntheses, Structures, and Nonlinear Optical Properties of Two Sulfides $Na_2In_2MS_6$ (M = Si, Ge). *Inorg. Chem.* **2016**, *55*, 1480–1485. [CrossRef] [PubMed]

57. Wu, Q.; Meng, X.G.; Zhong, C.; Chen, X.G.; Qin, J.G. $Rb_2CdBr_2I_2$: A New IR Nonlinear Optical Material with a Large Laser Damage Threshold. *J. Am. Chem. Soc.* **2014**, *136*, 5683–5686. [CrossRef] [PubMed]

58. Zhang, G.; Li, Y.J.; Jiang, K.; Zeng, H.Y.; Liu, T.; Chen, X.G.; Qin, J.G.; Lin, Z.S.; Fu, P.Z.; Wu, Y.C.; et al. A New Mixed Halide, $Cs_2HgI_2Cl_2$: Molecular Engineering for a New Nonlinear Optical Material in the Infrared Region. *J. Am. Chem. Soc.* **2012**, *134*, 14818–14822. [CrossRef] [PubMed]

59. Haynes, A.S.; Saouma, F.O.; Otieno, C.O.; Clark, D.J.; Shoemaker, D.P.; Jang, J.I.; Kanatzidis, M.G. Phase-Change Behavior and Nonlinear Optical Second and Third Harmonic Generation of The One-Dimensional $K_{(1-x)}Cs_xPSe_6$ and Metastable β-$CsPSe_6$. *Chem. Mater.* **2015**, *27*, 1837–1846. [CrossRef]

60. Wu, K.; Yang, Z.H.; Pan, S.L. $Na_4MgM_2Se_6$ (M = Si, Ge): The First Noncentrosymmetric Compounds with Special Ethane-Like $[M_2Se_6]^{6-}$ Units Exhibiting Large Laser-Damage Thresholds. *Inorg. Chem.* **2015**, *54*, 10108–10110. [CrossRef] [PubMed]

61. Wu, K.; Yang, Z.H.; Pan, S.L. $Na_2Hg_3M_2S_8$ (M = Si, Ge, and Sn): New Infrared Nonlinear Optical Materials with Strong Second Harmonic Generation Effects and High Laser-Damage Thresholds. *Chem. Mater.* **2016**, *28*, 2795–2801. [CrossRef]

62. Wu, K.; Yang, Z.H.; Pan, S.L. Na_2BaMQ_4 (M =Ge, Sn; Q=S, Se): Infrared Nonlinear Optical Materials with Excellent Performances and that Undergo Structural Transformations. *Angew. Chem. Int. Ed.* **2016**, *128*, 6825–6827. [CrossRef]

63. Zhen, N.; Nian, L.Y.; Li, G.M.; Wu, K.; Pan, S.L. A High Laser Damage Threshold and a Good Second-Harmonic Generation Response in a New Infrared NLO Material: $LiSm_3SiS_7$. *Crystals* **2016**, *6*, 121. [CrossRef]

64. Li, G.M.; Wu, K.; Liu, Q.; Yang, Z.H.; Pan, S.L. $Na_2ZnGe_2S_6$: A New Infrared Nonlinear Optical Material with Good Balance between Large Second-Harmonic Generation Response and High Laser Damage Threshold. *J. Am. Chem. Soc.* **2016**, *138*, 7422–7428. [CrossRef] [PubMed]

65. Pan, M.Y.; Ma, Z.J.; Liu, X.C.; Xia, S.Q.; Tao, X.T.; Wu, K.C. $Ba_4AgGa_5Pn_8$ (Pn = P, As): New Pnictide-Based Compounds with Nonlinear Optical Potential. *J. Mater. Chem. C* **2015**, *3*, 9695–9700. [CrossRef]

66. Kuo, S.M.; Chang, Y.M.; Chung, I.; Jang, J.I.; Her, B.H.; Yang, S.H.; Ketterson, J.B.; Kanatzidis, M.G.; Hsu, K.F. New Metal Chalcogenides $Ba_4CuGa_5Q_{12}$ (Q = S, Se) Displaying Strong Infrared Nonlinear Optical Response. *Chem. Mater.* **2013**, *25*, 2427–2433. [CrossRef]

67. Bera, T.K.; Jang, J.I.; Song, J.H.; Malliakas, C.D.; Freeman, A.J.; Ketterson, J.B.; Kanatzidis, M.G. Soluble Semiconductors $AAsSe_2$ (A = Li, Na) with a Direct-Band-Gap and Strong Second Harmonic Generation: A Combined Experimental and Theoretical Study. *J. Am. Chem. Soc.* **2010**, *132*, 3484–3495. [CrossRef] [PubMed]

68. Liao, J.H.; Marking, G.M.; Hsu, K.F.; Matsushita, Y.; Ewbank, M.D.; Borwick, R.; Cunningham, P.; Rosker, M.J.; Kanatzidis, M.G. α- and β-$A_2Hg_3M_2S_8$ (A = K, Rb; M= Ge, Sn): Polar Quaternary Chalcogenides with Strong Nonlinear Optical Response. *J. Am. Chem. Soc.* **2003**, *125*, 9484–9493. [CrossRef] [PubMed]

69. Parthé, E. *Crystal Chemistry of Tetrahedral Structures*; Gordon and Breach Science: New York, NY, USA, 1964.

70. Aitken, J.A.; Larson, P.; Mahanti, S.D.; Kanatzidis, M.G. Li_2PbGeS_4 and Li_2EuGeS_4: Polar Chalcopyrites with a Severe Tetragonal Compression. *Chem. Mater.* **2001**, *13*, 4714–4721. [CrossRef]

71. Bai, L.; Lin, Z.S.; Wang, Z.Z.; Chen, C.T. Mechanism of Linear and Nonlinear Optical Effects of Chalcopyrites $LiGaX_2$ (X = S, Se and Te) Crystals. *J. Appl. Phys.* **2008**, *103*, 083111. [CrossRef]

72. Kim, Y.; Seo, I.S.; Martin, S.W.; Baek, J.; Shiv Halasyamani, P.; Arumugam, N.; Steinfink, H. Characterization of New Infrared Nonlinear Optical Material with High Laser Damage Threshold, $Li_2Ga_2GeS_6$. *Chem. Mater.* **2008**, *20*, 6048–6052. [CrossRef]

73. Shi, Y.F.; Chen, Y.K.; Chen, M.C.; Wu, L.M.; Lin, H.; Zhou, L.J.; Chen, L. Strongest Second Harmonic Generation in the Polar R_3MTQ_7 Family: Atomic Distribution Induced Nonlinear Optical Cooperation. *Chem. Mater.* **2015**, *27*, 1876–1884. [CrossRef]

74. Brant, J.A.; Clark, D.J.; Kim, Y.S.; Jang, J.I.; Zhang, J.H.; Aitken, J.A. Li_2CdGeS_4, A Diamond-Like Semiconductor with Strong Second-Order Optical Nonlinearity in the Infrared and Exceptional Laser Damage Threshold. *Chem. Mater.* **2014**, *26*, 3045–3048. [CrossRef]

75. Brant, J.A.; Clark, D.J.; Kim, Y.S.; Jang, J.I.; Weiland, A.; Aitken, J.A. Outstanding Laser Damage Threshold in Li_2MnGeS_4 and Tunable Optical Nonlinearity in Diamond-Like Semiconductors. *Inorg. Chem.* **2015**, *4*, 2809–2819. [CrossRef] [PubMed]

76. Rosmus, K.A.; Brant, J.A.; Wisneski, S.D.; Clark, D.J.; Kim, Y.S.; Jang, J.I.; Brunetta, C.D.; Zhang, J.H.; Srnec, M.N.; Aitken, J.A. Optical Nonlinearity in Cu_2CdSnS_4 and α/β-Cu_2ZnSiS_4: Diamond-like Semiconductors with High Laser-Damage Thresholds. *Inorg. Chem.* **2014**, *53*, 7809–7811. [CrossRef] [PubMed]

77. Brant, J.A.; Cruz, C.D.; Yao, J.L.; Douvalis, A.P.; Bakas, T.; Sorescu, M.; Aitken, J.A. Field-Induced Spin-Flop in Antiferromagnetic Semiconductors with Commensurate and Incommensurate Magnetic Structures: Li_2FeGeS_4 (LIGS) and Li_2FeSnS_4 (LITS). *Inorg. Chem.* **2014**, *53*, 12265–12274. [CrossRef] [PubMed]

78. Zhang, J.H.; Clark, D.J.; Brant, J.A.; Sinagra, C.W.; Kim, Y.S.; Jang, J.I.; Aitken, J.A. Infrared Nonlinear Optical Properties of Lithium-Containing Diamond-Like Semiconductors $Li_2ZnGeSe_4$ and $Li_2ZnSnSe_4$. *Dalton Trans.* **2015**, *44*, 11212–11222. [CrossRef] [PubMed]

79. Devlin, K.P.; Glaid, A.J.; Brant, J.A.; Zhang, J.H.; Srnec, M.N.; Clark, D.J.; Kim, Y.S.; Jang, J.I.; Daley, K.R.; Moreau, M.A.; et al. Polymorphism and Second Harmonic Generation in a Novel Diamond-Like Semiconductor: Li_2MnSnS_4. *J. Solid State Chem.* **2015**, *231*, 256–266. [CrossRef]

80. Lekse, J.W.; Leverett, B.M.; Lake, C.H.; Aitken, J.A. Synthesis, physicochemical characterization and crystallographic twinning of Li_2ZnSnS_4. *J. Solid State Chem.* **2008**, *181*, 3217–3222. [CrossRef]

81. Rotermund, F.; Petrov, V.; Noack, F. Difference-Frequency Generation of Intense Femtosecond Pulses in the Mid-IR (4–12 μm) Using $HgGa_2S_4$ and $AgGaS_2$. *Opt. Commun.* **2000**, *185*, 177–183. [CrossRef]

82. Wu, K.; Su, X.; Pan, S.L.; Yang, Z.H. Synthesis and Characterization of Mid-Infrared Transparency Compounds: Acentric $BaHgS_2$ and Centric $Ba_8Hg_4S_5Se_7$. *Inorg. Chem.* **2015**, *54*, 2772–2779. [CrossRef] [PubMed]

83. Li, C.; Yin, W.L.; Gong, P.F.; Li, X.X.; Zhou, M.L.; Mar, A.; Lin, Z.S.; Yao, J.Y.; Wu, Y.C.; Chen, C.T. Trigonal Planar $[HgSe_3]^{4-}$ Unit: A New Kind of Basic Functional Group in IR Nonlinear Optical Materials with Large Susceptibility and Physicochemical Stability. *J. Am. Chem. Soc.* **2016**, *138*, 6135–6138. [CrossRef] [PubMed]

84. Wu, K.; Yang, Z.H.; Pan, S.L. The First Quaternary Diamond-like Semiconductor with 10-membered LiS_4 Rings Exhibiting Excellent Nonlinear Optical Performances. *Chem. Commun.* **2017**, *53*, 3010–3013. [CrossRef] [PubMed]

85. I Brese, N.E.; O'Keeffe, M. Bond-Valence Parameters for Solid. *Acta Crystallogr. B* **1991**, *47*, 192–197. [CrossRef]

86. Brown, I.D.; Altermatt, D. Bond-Valence Parameters Obtained from a Systematic Analysis of the Inorganic Crystal Structure Database. *Acta Crystallogr. B* **1985**, *41*, 244–247. [CrossRef]

87. Brown, I.D. *The Chemical Bond in Inorganic Chemistry: The Bond Valence Model*, 1st ed.; Oxford University Press: Oxford, UK, 2002.

88. Preiser, C.; Losel, J.; Brown, I.D.; Kunz, M.; Skowron, A. Long Range Coulomb Forces and Localized Bonds. *Acta Crystallogr. B* **1999**, *55*, 69–711. [CrossRef]

89. Salinas-Sanchez, A.; Garcia-Munoz, J.L.; Rodriguez-Carvajal, J.; Saez-Puche, R.; Martinez, J.L. Structural Characterization of R_2BaCuO_5 (R = Y, Lu, Yb, Tm, Er, Ho, Dy, Gd, Eu and Sm) Oxides by X-ray and Neutron Diffraction. *J. Solid State Chem.* **1992**, *100*, 201–211. [CrossRef]

90. Chen, S.; Walsh, A.; Luo, Y.; Yang, J.H.; Gong, X.G.; Wei, S.H. Wurtzite-Derived Polytypes of Kesterite and Stannite Quaternary Chalcogenide Semiconductors. *Phys. Rev. B* **2010**, *82*, 195203. [CrossRef]

91. Bhar, G.C.; Smith, R.C. Optical Properties of II–IV–V_2 and I–III–VI_2 crystals with Particular Reference to Transmission Limits. *Phys. Status Solidi A* **1972**, *13*, 157–168. [CrossRef]

92. Sheldrick, G.M. *SHELXTL*, version 6.14; Bruker Analytical X-ray Instruments, Inc.: Madison, WI, USA, 2008.

93. Spek, A.L. Single-Crystal Structure Validation with the Program PLATON. *J. Appl. Crystallogr.* **2003**, *36*, 7–13. [CrossRef]

© 2017 by the authors. Licensee MDPI, Basel, Switzerland. This article is an open access article distributed under the terms and conditions of the Creative Commons Attribution (CC BY) license (http://creativecommons.org/licenses/by/4.0/).

Article

Microstructure and Dielectric Properties of PTFE-Based Composites Filled by Micron/Submicron-Blended CCTO

Chao Xie, Fei Liang *, Min Ma, Xizi Chen, Wenzhong Lu and Yunxiang Jia

School of Optical and Electronic Information, Huazhong University of Science and Technology,
Wuhan 430074, China; chaoxhust@163.com (C.X.); lvmsg@hust.edu.cn (M.M.); 15671677452@163.com (X.C.);
lwz@hust.edu.cn (W.L.); 13429856225@163.com (Y.J.)
* Correspondence: liangfei@mail.hust.edu.cn

Academic Editor: Stevin Snellius Pramana
Received: 12 March 2017; Accepted: 25 April 2017; Published: 30 April 2017

Abstract: This paper investigated a polymer-based composite by homogeneously embedding calcium copper titanate ($CaCu_3Ti_4O_{12}$; CCTO) fillers into a polytetrafluoroethylene matrix. We observed the composite filled by CCTO powder at different sizes. The particle size effects of the CCTO filling, including single-size particle filling and co-blending filling, on the microstructure and dielectric properties of the composite were discussed. The dielectric performance of the composite was investigated within the frequency range of 100 Hz to 1 MHz. Results showed that the composite filled by micron/submicron-blended CCTO particles had the highest dielectric constant ($\varepsilon_r = 25.6$ at 100 Hz) and almost the same dielectric loss ($\tan\delta = 0.1$ at 100 Hz) as the composite filled by submicron CCTO particles at the same volume percentage content. We researched the theoretical reason of the high permittivity and low dielectric loss. We proved that it was effective in improving the dielectric property of the polymer-based composite by co-blending filling in this experiment.

Keywords: CCTO; polymers; dielectrics; permittivity

---

1. Introduction

Substrate materials have gained considerable attention because of their roles in the continuous development of the high-technology electronic industry [1]. Traditional polymer materials, such as polytetrafluoroethylene (PTFE) [2,3] and epoxy [4], are flexible and can be produced by a simple process. However, these materials cannot meet the developing trends of miniaturization and integration of electronic systems because of their low dielectric permittivity [5–7]. Ceramic/polymer composites, in which particles with high dielectric permittivity are used as fillers and polymers are used as the matrix, combine the merits of ceramics and polymers with high dielectric permittivity and excellent mechanical properties. These composites exhibit potential for various applications [8–11]. For instance, they are widely used in high charge-storage capacitors and high-speed integrated circuits, as well as other fields [12–14]. Materials containing calcium copper titanate ($CaCu_3Ti_4O_{12}$; CCTO) also show potential to be applied to embedded devices, etc. Compared with CCTO/polyvinylidene fluoride (PVDF) composites, which have been widely studied, our results show lower losses, which indicate these filled CCTO/PTFE materials can be applied to fields requiring high dielectric constants and low losses, such as substrate materials.

Calcium copper titanate ($CaCu_3Ti_4O_{12}$; CCTO) is one of the most remarkable ceramic materials because of its high dielectric permittivity, reaching as high as 10^4 to 10^5, its almost non-dependence on frequency up to 10 MHz, and its low thermal coefficient of dielectric permittivity from 100 K to 600 K [15,16]. To develop applications for CCTO, scholars aim to minimize the high dielectric loss [17–19]. Simultaneously, attempts have been made to explore the possibility of obtaining high dielectric permittivity composites for potential electronic components by embedding CCTO particles into polymer [20–23]. Dang [24] utilized in situ polymerization to disperse CCTO powders into a polyimide matrix and obtained a composite film with a dielectric permittivity of 49 when the volume fraction was 40 vol % at 100 Hz. Chi [25] reported that a relatively high dielectric permittivity, low loss, and low conductivity were simultaneously achieved in nano-sized CCTO/PI films. Compared with the micro-sized CCTO/PI film with a 10 vol % concentration of micro-sized CCTO, the dielectric permittivity of the nano-sized CCTO/PI film with a 3 vol % concentration of nano-sized CCTO increased by 16%. Yang [26] investigated the effect of nano- and micro-sized CCTO on the CCTO/PVDF composite. The effective dielectric constant (ε_r) of the composite containing 40 vol % nano-sized CCTO filler is more than 10^6 at 10^2 Hz and room temperature. This value is substantially higher than the dielectric constant of the composite containing micro-sized CCTO, with a ε_r value of 35.7 (at 40 vol %).

When the single-particle size of the CCTO powder is filled at high amounts, the CCTO particles cannot be totally encased by the PTFE matrix. The particles agglomerate to form voids, which are difficult to be filled by the PTFE matrix, thereby decreasing the dielectric constant of the composite. The volume percentage fraction of the ceramic filler should be less than 40% to reduce the porosity and preserve the mechanical properties. In our experiment, we fabricated a CCTO/PTFE composite film with 30 vol % filler. To solve these limitations, researchers have investigated double-particle-sized hybrids characterized by high density [27]. In this paper, we synthesized the submicro-sized CCTO powder by the oxalate co-precipitation method and the micro-sized CCTO powder by the solid-state reaction method. Both CCTO powders were blended at a proportion of 1:1 and the mixture was filled in the PTFE matrix. The microstructure and dielectric properties of the CCTO/PTFE composite with different particle sizes were investigated and compared. Moreover, the dielectric mechanism of the PTFE/CCTO composite is discussed in detail.

2. Experimental Section

Submicro-sized CCTO precursor powders were obtained through co-precipitation. The metal chlorides ($CaCl_2$, $TiCl_3$, and $CuCl_2$–$2H_2O$) were dissolved in water and added as precipitation agents to ethanol-containing oxalic acid. The precursors were calcined in air at 750 °C, 800 °C, and 850 °C for 10 h to obtain the oxide powders [28]. Micro-sized CCTO powder was synthesized by the solid-state reaction method [29]. Calcium ($CaCO_3$, 99.9%), copper oxide (CuO, 99.9%), and titanium dioxide (TiO_2, 99.9%), were purchased from Sigma-Aldrich (Chemical Reagent Co., Ltd., Shanghai, China) and used as raw materials. High-purity metal oxides were weighed based on the stoichiometric ratio and calcined at 950 °C for 10 h to obtain CCTO particles. Both CCTO powders were blended at a proportion of 1:1 and the mixture was ground with a titanate coupling agent to enhance the interface between the matrix and the filler. Emulsion polymerization was employed to disperse the CCTO powder into the PTFE emulsion (Shanghai 3F New Material Co., Ltd., Shanghai, China) with magnetic agitation at 90 °C [30]. The obtained dry mixture was smashed into a powder after heat treatment in a muffle furnace at 270 °C. The powder was pressed to obtain tablet samples approximately 18 mm in diameter and 1 mm in thickness. Cylindrically-shaped samples were prepared for microwave frequency measurements. In addition, samples were calcined at a low temperature of 370 °C to obtain the composite.

The sample structure was examined by X-ray diffraction (XRD; XRD-7000, Shimadzu, Kyoto, Japan) with Cu Kα_1 radiation (λ = 0.154056 nm) over the 10°–80° 2θ range. The microstructures of the freshly-fractured cross-section of the PTFE/CCTO composites were examined using a field-emission scanning electron microscope (FESEM, Sirion 200, FEI, Eindhoven, Netherlands), and we used Image-Pro Plus 6.0 to obtain the porosity of three different particle-sized samples by calculating the area of the black part in the SEM micrographs. The thermal analysis system used in this study was TGA-DSC (METTLER TOLEDO, TGA/DSC-1). The dielectric properties of composite were measured using an impedance analyzer (Agilent 4294A, Agilent Technologies, Santa Clara, CA, USA) at room temperature in the frequency range of 100 Hz to 1 MHz. A thin layer of silver was painted on both sides of the tablet samples before measurement.

3. Results and Discussion

Figure 1 presents the thermal gravimetric analysis (TGA) of the CCTO precursor. The thermal process is mainly divided into three stages. The weightlessness of the CCTO precursor is about 8.71% at the first stage from 50 °C to 200 °C. The corresponding DSC curve appears to be a weak endothermic peak which is attributed to water evaporation. At the second stage, from 200 °C to 300 °C, the weightlessness of the CCTO precursor is about 45.07% and the corresponding DSC curve appears to have a sharp exothermic peak which is attributed to the combustion decomposition of organic matter, such as oxalate and nitrate. At the third stage, from 560 °C to 700 °C, as the weightlessness of the CCTO precursor is about 2.17%, and the corresponding DSC curve appears to have a weak exothermic peak which may correspond to CCTO crystallization. Figure 2 shows the XRD pattern of CCTO powders calcined at 750 °C, 800 °C, and 850 °C. CCTO appears as the major phase, and small amounts of $CaTiO_3$ and CuO are also present in the three samples; however, TiO_2 is only present in the sample calcined at 750 °C.

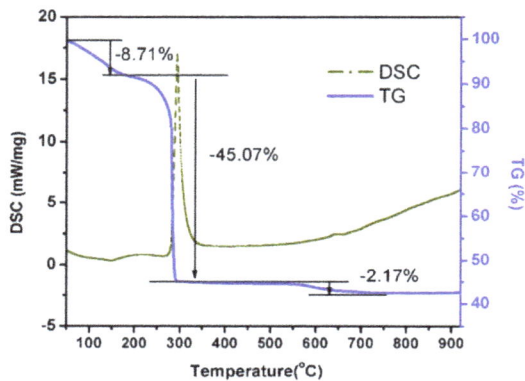

Figure 1. TG–DSC curve of the submicron CCTO precursor.

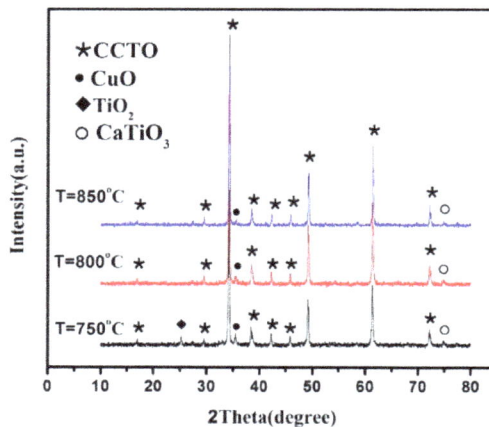

Figure 2. XRD patterns of submicron CCTO at different calcination temperatures.

Figure 3 shows the scanning electron microscopy (SEM) images of the CCTO powder calcined at different temperatures: (a) 750 °C, (b) 800 °C, and (c) 850 °C. Some spherical and rod-like particles appear in the CCTO powders calcined at 750 °C. The CCTO powders with sizes of about 100–200 nm are uniform and spherical at the calcination temperature of 800 °C. When the temperature rises to 850 °C, the CCTO particles grow further and their sizes increase to about 300 nm to 500 nm. According to the XRD and SEM results, the synthesis temperature of the submicron CCTO powder was determined to be 800 °C, in order to obtain a smaller and homogeneous submicron grain and reflect the effect of particle size on the dielectric properties of the composites as much as possible.

Figure 3. SEM micrograph of the CCTO powder calcined at different temperatures: (a) 750 °C, (b) 800 °C and (c) 850 °C.

Figure 4 shows SEM micrographs of the CCTO powder at different particle sizes. The particle size of the micron CCTO powder prepared by the solid-state reaction method is not uniform, ranging from 2 μm to 10 μm. CCTO powders prepared by the co-precipitation method appear to have spherical particles with uniform sizes. However, these particles agglomerated to form a cluster with some voids existing in them. In the micron/submicron co-blended CCTO powder, micron-sized particles are uniformly distributed while submicron particles fill in the gaps between the micron-sized particles. As shown in Figure 4, the micron/submicron co-blended CCTO powder has a significantly higher filling density than the other two groups with a single particle size.

Figure 4. SEM micrograph of CCTO powder with different particle sizes: (a) micron, (b) submicron, and (c) micron/submicron co-blending.

Figure 5 shows the SEM micrograph of the PTFE-based composite filled by 30 vol % CCTO powder with different particle sizes. Micron CCTO powders were uniformly coated by the PTFE matrix, while submicron CCTO powders were not completely coated by PTFE because of their agglomeration. Comparing the two kinds of composites, the composite filled by the micron/submicron co-blended CCTO powder showed the highest density because the submicron particles filled the gaps not only among micron-sized CCTO particles, but also between micron-sized CCTO particles and the PTFE matrix to minimize the surface energy of the system. According to our statistics, the porosity of the composites filled with micron, submicron, and micron/submicron co-blending CCTO is, respectively, 19.43%, 13.98%, and 11.05%. Lower porosity leads to higher dielectric permittivity [17,31], as shown in Figure 7. In addition, the surface area-volume ratio of submicron CCTO powder is 88,737 cm^2/cm^3, much larger than 13,425 cm^2/cm^3 of micron CCTO powder. The stronger interfacial polarization mostly caused by the large specific surface area also contributes to the high permittivity [21,26].

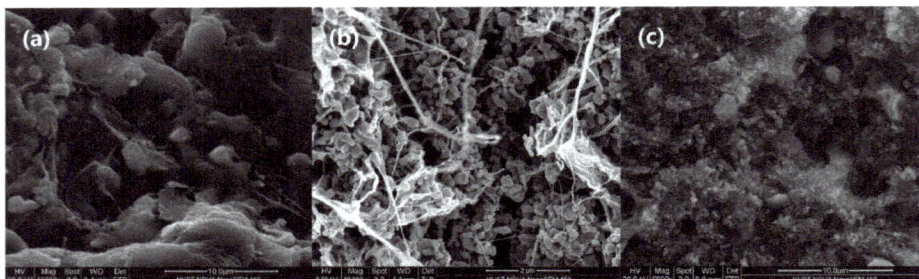

Figure 5. SEM micrograph of the PTFE-based composite filled by different CCTO particle sizes: (**a**) micron CCTO filling; (**b**) submicron CCTO filling; and (**c**) micron/submicron co-blending CCTO filling.

Figure 6 shows the XRD patterns of the PTFE-based composite filled with 30 vol % CCTO powder at different sizes. All three composite materials have the characteristic peak of CCTO and PTFE. The composite filled by micron CCTO has an obviously higher CCTO peak intensity than the other two groups. This is because the growth of micron CCTO ceramic particles is more complete, with the particles having better crystallinity. In addition, the great specific surface area of submicron CCTO results in diminishment of its peak intensity. Conversely, the effect results in the uniform distribution of micron-sized CCTO particles in the PTFE matrix and less damage of the PTFE chain. Therefore, the PTFE characteristic peak at $2\theta = 18.2°$ is stronger than that of the other two groups. Furthermore, Figure 6 shows that the PTFE characteristic peak in the composite filled by micron and submicron CCTO particles is the weakest and, together with Figure 5, they indicate that the PTFE chain destruction is stronger than that of the other two groups, and the composite materials filled by micron and submicron CCTO particles exhibit the highest density.

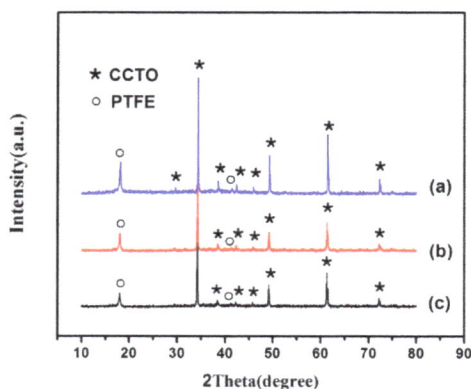

Figure 6. XRD patterns of the PTFE-based composite filled by different CCTO particle sizes: (**a**) micron CCTO filling; (**b**) submicron CCTO filling; and (**c**) micron/submicron-blended CCTO filling.

Figure 7 shows the frequency dependence of the dielectric properties of the composites with different CCTO particle sizes. All dielectric constants of the three composites show good frequency stability. Moreover, the composite filled by the double-particle-sized CCTO powder has a higher dielectric constant, which is up to 25.6 at a frequency of 100 Hz, than the other two composites. This is attributed to its stronger interfacial polarization and higher density. The dielectric constant of the composite filled by sub-micron CCTO particles is slightly higher than that of the composite filled by micron CCTO particles because a large number of voids are caused by the agglomeration of submicron CCTO particles. Additionally, the grain size effects contribute to the higher dielectric properties of the composite [32,33]. Figure 7 shows that the dielectric loss of the three composites decreased rapidly with increasing frequency at 100 Hz–1 kHz, but increased when the frequency increased to nearly 10^6 Hz. The dielectric loss of submicron and co-filled composites change more obviously than the micron-sized composite because of their greater specific surface area and stronger interfacial polarization.

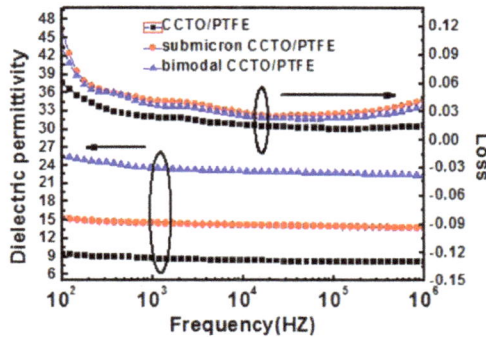

Figure 7. Frequency dependence of the dielectric properties of the composites with various CCTO particles: (**a**) the relative permittivity and (**b**) dielectric loss.

4. Conclusions

In this paper, PTFE was filled by micron, submicron, and micron/submicron-blended CCTO powder at three different sizes. The effects of the filler particle sizes on the microstructure and dielectric properties were discussed. The research results showed that the composite filled by micron/submicron-blended CCTO particles had the highest density and dielectric constant with the same volume percentage content. The dielectric loss of the three composites decreased with an increase in the frequency. The composite filled by micron/submicron CCTO particles had a higher dielectric constant and almost the same dielectric loss as the composite filled by submicron CCTO particles, which may be attributed to the higher density of the former.

Acknowledgments: The authors would like to acknowledge financial support from the National Natural Science Foundation of China through grant No. 61172004 and the Fundamental Research Funds for the Central Universities, No. 201509.

Author Contributions: Chao Xie and Fei Liang conceived and designed the experiments; Chao Xie, Min Ma, and Xizi Chen performed the experiments; and Wenzhong Lu and Yunxiang Jia analyzed the data. All authors participated in the research, analysis, and edition of the manuscript.

Conflicts of Interest: The authors declare no conflict of interest.

References

1. Youngs, I.J.; Stevens, G.C.; Voughan, A.S. Trends in dielectrics research: An international review from 1980 to 2004. *J. Phys. D Appl. Phys.* **2006**, *39*, 1267. [CrossRef]
2. Murali, K.P.; Rajesh, S.; Prakash, O.; Kulkarni, A.R.; Ratheesh, R. Comparison of alumina and magnesia filled PTFE composites for microwave substrate applications. *Mater. Chem. Phys.* **2009**, *113*, 290. [CrossRef]
3. Liang, F.; Zhang, L.; Lu, W.Z.; Wan, Q.X.; Fan, G.F. Dielectric performance of polymer-based composites containing core-shell Ag@TiO$_2$ nanoparticle fillers. *Appl. Phys. Lett.* **2016**, *108*, 072902. [CrossRef]
4. Prakash, B.S.; Varma, K.B.R. Dielectric behavior of CCTO/epoxy and Al-CCTO/epoxy composites. *Compos. Sci. Technol.* **2007**, *67*, 2363–2368. [CrossRef]
5. Dang, Z.M.; Yuan, J.K.; Zha, J.W.; Zhou, T.; Li, S.T.; Hu, G.H. Fundamentals, processes and applications of high-permittivity polymer—Matrix composites. *Prog. Mater. Sci.* **2012**, *57*, 660. [CrossRef]
6. Dang, Z.M.; Yuan, J.K.; Yao, S.H.; Liao, R.J. Flexible nanodielectric materials with high permittivity for power energy storage. *Adv. Mater.* **2013**, *25*, 6334–6365. [CrossRef] [PubMed]
7. Sebastian, M.T.; Jantunen, H. Polymer-ceramic composites of 0-3 connectivity for circuits in electronics: A review. *Int. J. Appl. Ceram. Technol.* **2010**, *7*, 415–434. [CrossRef]
8. Subodh, G.; Deepu, V.; Mohanan, P.; Sebastian, M.T. Dielectric response of high permittivity polymer ceramic composite with low loss tangent. *Appl. Phys. Lett.* **2009**, *95*, 062903. [CrossRef]
9. Singh, P.; Borkar, H.; Singh, B.P.; Singh, V.N.; Kumar, A. Ferroelectric polymer-ceramic composite thick films for energy storage applications. *AIP Adv.* **2014**, *4*, 087117. [CrossRef]
10. Pela'iz-Barranco, A. Dielectric relaxation and electrical conductivity in ferroelectric ceramic/polymer composites around the glass transition. *Appl. Phys. Lett.* **2012**, *100*, 212903. [CrossRef]
11. Lee, H.J.; Zhang, S.J.; Meyer, R.J., Jr.; Sherlock, N.P.; Shrout, T.R. Characterization of piezoelectric ceramics and 1-3 composites for high power transducers. *Appl. Phys. Lett.* **2012**, *101*, 032902. [CrossRef] [PubMed]
12. Lin, Y.-H.; Cai, J.; Li, M.; Nan, C.-W.; He, J. Grain boundary behavior in varistor-capacitor TiO$_2$-rich CaCu$_3$Ti$_4$O$_{12}$ ceramics. *J. Appl. Phys.* **2008**, *103*, 74111. [CrossRef]
13. Li, W.; Schwartz, R.W. Ac conductivity relaxation processes in CaCu$_3$Ti$_4$O$_{12}$ ceramics: Grain boundary and domain boundary effects. *Appl. Phys. Lett.* **2006**, *89*, 242906. [CrossRef]
14. Amaral, F.; Rubinger, C.P.L.; Henry, F.; Costa, L.C.; Valente, M.A.; Barros-Timmons, A. Dielectric properties of polystyrene-CCTO composite. *J. Non-Cryst. Solids* **2008**, *354*, 5321–5322. [CrossRef]
15. Subramanian, M.A.; Li, D.; Duan, N.; Reisner, B.A.; Sleight, A.W. High dielectric constant in ACu$_3$Ti$_4$O$_{12}$ and ACu$_3$Ti$_3$FeO$_{12}$ phases. *J. Solid State Chem.* **2000**, *151*, 323–325. [CrossRef]
16. Homes, C.C.; Vogt, T.; Shapiro, S.M. Optical Response of High-Dielectric-Constant Perovskite-Related Oxide. *Science* **2001**, *293*, 673. [CrossRef] [PubMed]
17. Liu, P.; Lai, Y.M.; Zeng, Y.M.; Wu, S.; Huang, Z.H.; Han, J. Influence of sintering conditions on microstructure and electrical properties of CaCu$_3$Ti$_4$O$_{12}$ (CCTO) ceramics. *J. Alloys Compd.* **2015**, *650*, 59–64. [CrossRef]
18. De Almeida-Didry, S.; Autret, C.; Lucas, A.; Honstettre, C.; Pacreau, F.; Gervais, F. Leading role of grain boundaries in colossal permittivity of doped and undoped CCTO. *J. Eur. Ceram. Soc.* **2014**, *34*, 3649–3654. [CrossRef]
19. Tang, H.; Zhou, Z.; Bowland, C.C.; Sodano, H.A. Synthesis of calcium copper titanate (CaCu$_3$Ti$_4$O$_{12}$) nanowires with insulating SiO$_2$ barrier for low loss high dielectric constant nanocomposites. *Nano Energy* **2015**, *17*, 302–307. [CrossRef]
20. Yang, Y.; Zhu, B.P.; Lu, Z.H.; Wang, Z.Y.; Fei, C.L.; Yin, D.; Xiong, R.; Shi, J.; Chi, Q.G.; Lei, Q.Q. Polyimide/nanosized CaCu$_3$Ti$_4$O$_{12}$ functional hybrid films with high dielectric permittivity. *Appl. Phys. Lett.* **2013**, *102*, 042904. [CrossRef]
21. Yang, Y.; Sun, H.L.; Yin, D.; Lu, Z.H.; Wei, J.H.; Xiong, R.; Shi, J.; Wang, Z.Y.; Liu, Z.Y.; Lei, Q.Q. High performance of polyimide/CaCu$_3$Ti$_4$O$_{12}$@Ag hybrid films with enhanced dielectric permittivity and low dielectric loss. *J. Mater. Chem. A* **2015**, *3*, 4916. [CrossRef]
22. Gao, L.; Wang, X.; Chen, Y.; Chi, Q.G.; Lei, Q.Q. Ni-coated CaCu$_3$Ti$_4$O$_{12}$/low density polyethylene composite material with ultra-high dielectric permittivity. *AIP Adv.* **2015**, *5*, 087183. [CrossRef]
23. Arbatti, M.; Shan, X.B.; Cheng, Z.Y. Ceramic–Polymer Composites with High Dielectric Constant. *Adv. Mater.* **2007**, *19*, 1369–1372. [CrossRef]

24. Dang, Z.M.; Zhou, T.; Yao, S.H.; Yuan, J.K.; Zha, J.W.; Song, H.T. Advanced Calcium Copper Titanate polyimide. *Adv. Mater.* **2009**, *21*, 2077–2082. [CrossRef]
25. Chi, Q.G.; Sun, J.; Zhang, C.H.; Liu, G.; Lin, J.Q.; Wang, Y.N.; Wang, X.; Lei, Q.Q. Enhanced dielectric performance of amorphous calcium copper titanate/polyimide hybrid film. *J. Mater. Chem. C* **2014**, *2*, 172–177. [CrossRef]
26. Yang, W.H.; Yu, S.H.; Sun, R.; Du, R.X. Nano- and microsize effect of CCTO fillers on the dielectric behavior of CCTO/PVDF composites. *Acta Mater.* **2011**, *59*, 5593–5602. [CrossRef]
27. Zha, J.W.; Zhu, Y.H.; Li, W.K.; Bai, J.B.; Dang, Z.M. Low dielectric permittivity and high thermal conductivity silicone rubber composites with micro-nano-sized particles. *Appl. Phys. Lett.* **2012**, *101*, 062905. [CrossRef]
28. Marchin, L.; Guillemet-Fritsch, S.; Durand, B. Soft chemistry synthesis of the perovskite $CaCu_3Ti_4O_{12}$. *Prog. Solid State Chem.* **2007**, *36*, 151–155. [CrossRef]
29. Ehrhardt, C.; Fettkenhauer, C.; Glenneberg, J.; Münchgesang, W.; Leipner, H.S.; Diestelhorst, M. A solution-based approach to composite dielectric films of surface functionalized $CaCu_3Ti_4O_{12}$ and P(VDF-HFP). *J. Mater. Chem. A* **2014**, *2*, 2266. [CrossRef]
30. Barbier, B.; Combettes, C.; Barbiera, B.; Guillemet-Fritschb, S.; Chartierc, T.; Rossignolc, F.; Rumeaud, A.; Lebeyd, T.; Dutardea, E. $CaCu_3Ti_4O_{12}$ ceramics from co-precipitation method: Dielectric properties of pellets and thick films. *J. Eur. Ceram. Soc.* **2009**, *29*, 731. [CrossRef]
31. Liu, J.; Gan, D.; Hu, C.; Kiene, M.; Ho, P.S.; Volksen, W.; Miller, R.D. Porosity effect on the dielectric constant and thermomechanical properties of organosilicate films. *Appl. Phys. Lett.* **2002**, *81*, 4180–4182. [CrossRef]
32. Lee, J.; Koh, J. Grain size effects on the dielectric properties of $CaCu_3Ti_4O_{12}$ ceramics for supercapacitor applications. *Ceram. Int.* **2015**, *41*, 10442–10447. [CrossRef]
33. Yu, V.; Dang, Z.; Zha, J. Micro-Nanosize Cofilled High Dielectric Permittivity Composites. In Proceedings of the IEEE 9th International Conference on the Properties and Applications of Dielectric Materials, Harbin, China, 19–23 July 2009; pp. 769–772.

© 2017 by the authors. Licensee MDPI, Basel, Switzerland. This article is an open access article distributed under the terms and conditions of the Creative Commons Attribution (CC BY) license (http://creativecommons.org/licenses/by/4.0/).

crystals

MDPI

Article

A Diagram of the Structure Evolution of Pb(Zn$_{1/3}$Nb$_{2/3}$) O$_3$-9%PbTiO$_3$ Relaxor Ferroelectric Crystals with Excellent Piezoelectric Properties

Hua Zhou [1,2], Tao Li [1,2,3], Nian Zhang [4], Manfang Mai [1,2], Mao Ye [1,2], Peng Lin [1], Chuanwei Huang [1], Xierong Zeng [1,2], Haitao Huang [3] and Shanming Ke [1,*]

[1]   College of Materials Science and Engineering and Shenzhen Key Laboratory of Special Functional Materials, Shenzhen University, Shenzhen 518060, China; zhouhua3612@163.com (H.Z.); litao@fjirsm.ac.cn (T.L.); mfmai1121@hotmail.com (M.M.); kfadn0125@gmail.com (M.Y.); lin.peng@szu.edu.cn (P.L.); cwhuang@szu.edu.cn (C.H.); zengxier@szu.edu.cn (X.Z.)
[2]   College of Optoelectronic Engineering and Key Laboratory of Optoelectronic Devices and Systems of Ministry of Education and Guangdong Province, Shenzhen University, Shenzhen 518060, China
[3]   Department of Applied Physics and Materials Research Centre, The Hong Kong Polytechnic University, Hung Hom, Kowloon, Hong Kong; aphhuang@polyu.edu.hk
[4]   Center for Excellence in Superconducting Electronics, Shanghai Institute of Microsystem and Information Technology, Chinese Academy of Sciences, 865 Changning Road, Shanghai 200050, China; zhangn@mail.sim.ac.cn
*   Correspondence: smke@szu.edu.cn; Tel.: +86-755-2653-4059

Academic Editor: Stevin Snellius Pramana
Received: 27 March 2017; Accepted: 1 May 2017; Published: 8 May 2017

Abstract: Piezoelectric properties are of significant importance to medical ultrasound, actuators, sensors, and countless other device applications. The mechanism of piezoelectric properties can be deeply understood in light of structure evolutions. In this paper, we report a diagram of the structure evolutions of Pb(Zn$_{1/3}$Nb$_{2/3}$)$_{0.91}$Ti$_{0.09}$O$_3$ (PZN-9PT) crystals with excellent piezoelectric properties among orthorhombic, tetragonal, and cubic phases, with a temperature increasing from room temperature to 220 °C. Through fitting the temperature-dependent XRD curves with Gauss and Lorenz functions, we obtained the evolutions of the content ratio of three kinds of phases (orthorhombic, tetragonal and cubic) and the lattice parameters of the PZN-9PT system with the changes of temperature. The XRD fitting results together with Raman and dielectric spectra show that the phase transitions of PZN-9PT are a typical continuous evolution process. Additionally, resonance and anti-resonance spectra show the excellent piezoelectric properties of these crystals, which probably originate from the nano twin domains, as demonstrated by TEM images. Of particular attention is that the thickness electromechanical coupling factor k_t is up to 72%.

Keywords: Pb(Zn$_{1/3}$Nb$_{2/3}$) O$_3$-9%PbTiO$_3$; relaxor ferroelectric crystals; structure phase transition; electromechanical coupling factor

1. Introduction

Since Kuwata et al. [1] first reported that the solid solutions of rhombohedral lead zinc niobate, Pb(Zn$_{1/3}$Nb$_{2/3}$)O$_3$ (PZN), and tetragonal lead titanate, PbTiO$_3$ (PT), have a morphotropic phase boundary (MPB) near 9 mol % PT, PZN-PT has attracted much exclusive attention due to its excellent and unique properties. Recently, Zhang et al. and Sun et al. [2–4] demonstrated that the Pb(Zn$_{1/3}$Nb$_{2/3}$)$_{0.91}$Ti$_{0.09}$O$_3$ (PZN-9PT) crystal exhibits an exceptionally large piezoelectric constant (d_{33} > 2000 pC/N) and an electromechanical coupling factor in longitudinal bar mode (k_{33} > 92%). These properties are expected to be applied in high-quality piezoelectric devices, such as medical ultrasonic transducers, actuators, and sonars, as replacements for conventional PZT ceramics [4–7]. In the PZN-9PT crystal system, different from the normal ferroelectric (e.g., PT) [8] with a long coherent length of order, the random distribution of B-ions (here, B represents Zn, Nb, and Ti atoms) prevents the formation of the long-range ferroelectric order [8,9]. Additionally, the charge and ionic radius differences are of great importance for the relaxor ferroelectric (RFE) properties, because they directly determine the degree of order in the B-ion sublattice and thus the coherence length of order [10,11]. Consequently, the phase diagram of the $Pb(B'B'')O_3$-$PbTiO_3$ systems are often more complex than the normal ferroelectrics. Through the X-ray diffraction (XRD), neutron diffraction, Raman spectrum, etc., many researchers have reported the evolution diagram of the $(1-x)$PZN-xPT phase transition with changes in x and temperature [12–15]. In spite of these intensive studies, their microscopic mechanisms, such as the ferroelectric phase transition in relaxor ferroelectrics, remain poorly understood. Moreover, the ratios among the different phases as the change of the system temperature always seem to be ignored by researchers, which should play an important role in understanding the phase transition mechanism.

In order to reveal the evolutions of the phase structure driven by the temperature, we have successfully grown PZN-9PT single crystals via controlled top-seeded solution growth (TSSG). Of particular importance is that the TSSG technique offers advantages in growing single crystals of good quality, low compositional segregation, and controllable morphology. A structure phase transition was observed by the in-situ variable temperature XRD, dielectric spectrum, and Raman spectra. The resonance and anti-resonance spectrum showed that the PZN-9PT crystal exhibits an excellent piezoelectric response, which probably originates from the nano twin structures in the PZN-9PT single crystal, as demonstrated by transmission electron microscopy (TEM).

2. Results and Discussion

2.1. Structure Phase Transition of PZN-9PT Crystals

It is universally acknowledged that PZN-9PT single crystals have five structure phases: monoclinic (M_C), rhombohedral (R), orthorhombic (O), tetragonal (T), and cubic (C), corresponding to their space groups: *Pm C*, *R3m*, *Amm2*, *P4mm*, and *Pm-3m*, respectively. PZN-9PT single crystals often possess the coexistent phase structures at room temperature. Ye et al. [16] and Chang et al. [17] reported the coexistence of R-phase and T-phase in PZN-9PT single crystals at room temperature. Later, Cox et al. [11,12] argued that the room temperature structure of PZN-9PT should be O-phase coexisting with T-phase. Generally speaking, for R-phase, the lattice parameters a = b = c, $\alpha = \beta = \gamma \neq 90°$ (but nearly equal to 90° for PZN-9PT system), and for O-phase, a $\neq$ b $\neq$ c, $\alpha = \beta = \gamma = 90°$. Based on these characteristics, we can distinguish the co-phase structures according to the XRD peak numbers of the {200} lattice plane. Figure 1a presents the survey of the variable-temperature XRD patterns of the powders made from the PZN-9PT relaxor ferroelectric single crystals. According to the changes of the XRD patterns of the PZN-9PT lattice planes: {100}, {110}, {200}, {112}, {220}, etc., it is inferred that structure phase transitions occur at about 80 °C and 150 °C for the powders of PZN-9PT. The color curve lines in Figure 1a, acquired using Poudrix software, show the calculated XRD patterns for the pure phases of PZN-9PT powders. The {002} plane families for the pure R-phase and the O-phase display one peak and three peaks, respectively, as shown in Figure 1b. The peak number of the experimental {200} result at room temperature is more than 3, indicating that PZN-9PT is not a pure O-phase at room temperature. Comparing the simulated results (shown by color lines in Figure 1a,b), it is speculated that the phase of single crystal PZN-9PT should be composed of O- and T-phases. This observation is well consistent with reports by Uesu et al. [18]. In addition, when the temperature rises to 150 °C or above, the peaks of {200} seem to merge into a single one, indicating the phase evolution from tetragonal/orthorhombic to cubic, as illustrated by the changes of blue and violet areas.

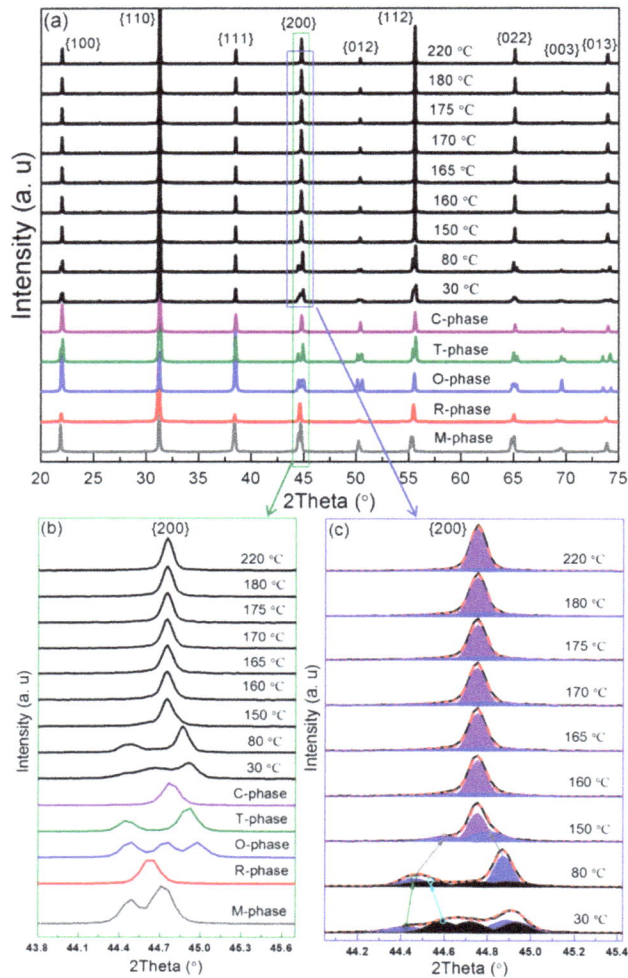

Figure 1. (**a**) The temperature dependent XRD patterns of PZN-9PT single crystals from 30 to 220 °C. Simulated XRD patterns (color lines at the bottom of Figure 1a) for pure monoclinic (M_C), rhombohedral (*R*), orthorhombic (*O*), tetragonal (*T*), and cubic (*C*) phase of PZN-9PT are also illustrated; (**b**,**c**) The amplified curves around {200} with fitting results.

According to the above discussions, it could be concluded that the O-phase and T-phase coexist in PZN-9PT at room temperature. When the temperature increases, O-phase transforms into T-phase firstly, and then transforms into C-phase gradually. In order to provide more detailed information of the phase transition processes, the Gauss and Lorenz functions (8:2) were used to fit the XRD data. Theoretically, for a pure orthorhombic structure, the areas of the three {200} peaks are equal as well as the full width at half maximum (FWHM). However, for a pure tetragonal structure with two {002} peaks, one peak area is twice as large as the other, while the FWHM is equal. The fitting parameters can then be set according to these quantitative relationships. The red dashed lines in Figure 1c show the fitting results, where the areas of the orthorhombic, tetragonal, and cubic phases are also illustrated. It is clear that the dominant phase at room temperature is the O-phase, as shown by the black areas in Figure 1c. The peak of the O-phase at 44.6° shifts to a lower angle with increasing temperatures, which implies that the *c*-axis becomes longer when the phase transits from orthorhombic to tetragonal phase. Moreover, the two {200} peaks of the T-phase tend to merge into a middle angle (shown by gray arrows in Figure 1c) with increasing temperatures, suggesting a shortened *c*-axis and larger *a*- and *b*-axes.

Based on the fitting data, the phase contents of PZN-9PT at different temperatures could be obtained and illustrated in Figure 2a. Four easily distinguishable regions can be identified as follows: Region I (O- and T-phases coexistence), Region II (T-phase dominant), Region III (T- and C-phases coexistence), and Region IV (C-phase dominant). It can be seen that the content of O-phase decreases with increasing temperature in Regions I and II, and subsequently becomes nearly zero in Regions III and IV; T-phase increases in Region I and then decreases gradually in the other regions. From the discussions above, one can conclude that the phase transition in PZN-9PT is a gradual evolution rather than a mutational change with increasing temperature. Due to the changes of phases, the dielectric properties of PZN-9PT display a temperature-dependent behavior accordingly, as shown in Figure 2b. Dielectric curves reveal only a weak frequency dispersion in Region I, while a strong dispersion could be observed with the coexistence of T- and C-phases. It is well known that the relaxor ferroelectrics are characterized by a broad frequency-dependent dielectric peak at radio frequencies, which is associated with the dimensional changes of polar nanoregions (PNRs). By combining 9% ferroelectric PT with PZN relaxors, the structure evolution becomes quite complicated. Upon cooling, some of the C-phase transforms into the T-phase and PNRs then occur in the crystal. In Region III, the C-phase is the dominant phase and PNRs grow in size, leading to a strong frequency dispersion of dielectric curves. The peak-like behaviors in Region III could be attributed to the configuration change of PNRs and/or microdomains. Moreover, in Region II, ferroelectric T-phase constitutes a majority and PNRs/microdomains grow into macrodomains, which is less sensitive to radio frequencies.

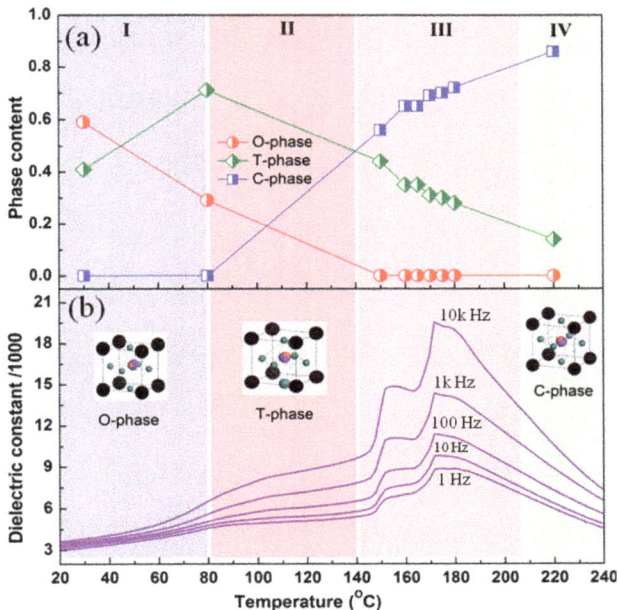

Figure 2. (**a**) The evolution of different phases in PZN-9PT. Four regions with different amounts of phases could be clearly distinguished; (**b**) The temperature dependence of the dielectric constant of PZN-9PT at selected frequencies (1, 10, 100, 1000, and 10,000 Hz).

Further evidence for the evolution of the phase structures in PZN-9PT can be found from the Raman spectra, which is very sensitive to the change in the structure, symmetry, and short-range order at nanoscale. As shown in Figure 3, six robust active modes could be observed, labeled as <Pb^{2+}>, C, σ and R, <Ti–O>, and <Nb–O>, respectively. These modes originate from the vibrations of the Pb^{2+} cation, coupling between Pb^{2+} and the neighboring entities, B–O bending and ion-covalent BO$_6$ vibration (B/Ti and Nb), stretching TiO$_6$ and NbO$_6$ modes, respectively. One can see that the σ, R, and <Ti–O> modes reveal a blue shift as well as a decrement of the peak intensity with increasing temperatures. These results are well consistent with the observations reported by Mishra et al. [19] and Cheng et al. [20]. The gradual shift of these modes also implies that the structure evolution of PZN-9PT is continuous rather than mutational, in agreement with the XRD results.

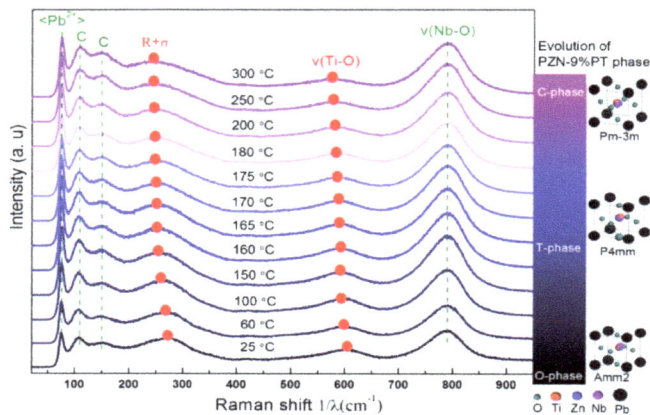

Figure 3. The temperature-dependent Raman spectra of PZN-9PT. Red solid circles show the shifts and intensity changes of the related Ti–O vibration peak. The right image shows a schematic of the structure evolution and corresponding atomic models.

2.2. Piezoelectric Properties of PZN-9PT Crystals

On account of the multi-phase coexistence, PZN-9PT possesses excellent piezoelectric, electromechanical, and ferroelectric properties. Figure 4a,b display the longitudinal mode impedance and phase angle, and the corresponding thickness mode impedance and phase angle at room temperature, respectively. Based on the resonance frequency (f_r) and anti-resonance frequency (f_a), the longitudinal (k_{33}) and thickness (k_t) electromechanical coupling factors are calculated to be 92% and 72%, respectively, by the following equations [21,22]:

$$k_{33}^2 = \frac{f_r}{f_a} \cot\left(\frac{\pi}{2}\frac{f_r}{f_a}\right) \tag{1}$$

and

$$k_t^2 = \frac{\pi}{2}\frac{f_r}{f_a} \tan\left(\frac{\pi}{2}\frac{f_a - f_r}{f_a}\right). \tag{2}$$

These results indicate that the PZN-9PT single crystals exhibit high electromechanical coupling properties. It is also worth noting that the k_t of TSSG-growth PZN-9PT is substantially superior to the modified Bridgman method [23]. The piezoelectric coefficient d_{33} is measured to be 2350 pC/N by a d_{33} meter. According to the resonance measurement, the d_{33} can be calculated to be 2240 pC/N which is slightly lower than the directly measured value.

The polarization-electric field (*P–E*) hysteresis loops of the PZN-9PT crystal were measured as a function of electric field and temperature at 2 Hz, which is shown in Figure 4c. It can be seen that the coercive field E_c decreases with the increase in temperature, but the statured remnant polarization P_r reaches at a maximum at T_{O-T} firstly and then decreases (Figure 4c), which results from the flexibility enhancement of domain switching and moving near the T_{O-T}. The statured remnant polarization P_r reaches 20 μC/cm², with a coercive electric field $E_c = 9$ kV/cm at room temperature (Figure 4d). This evolution between P_r and temperature indicate that the piezoelectric properties gradually become weak with the phase transition from the O-phase to the T-phase and to the C-phase. It is worth noting that the coercive field E_c of PZN-9PT single crystal is about 2~3 times larger than that of PMN–33 mol % PT, manifesting a more stable domain state in PZN–PT single crystals.

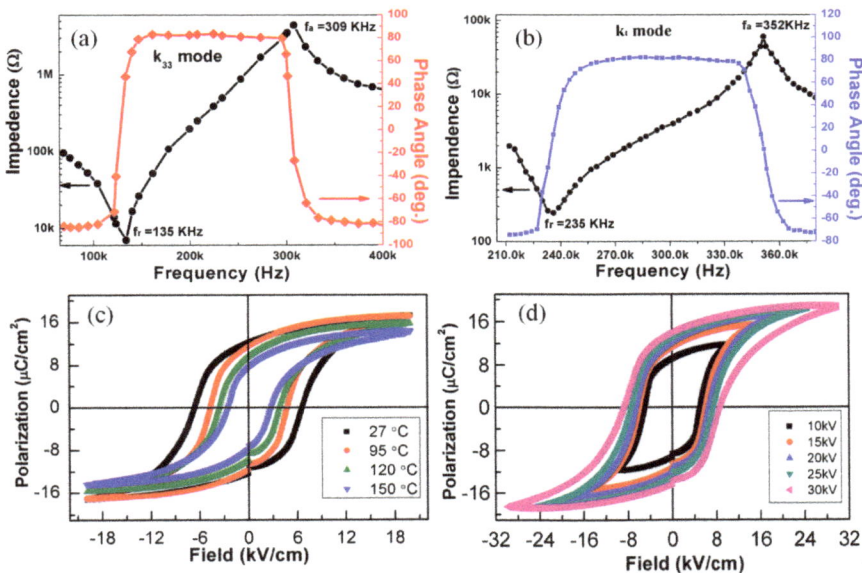

Figure 4. (**a**) Longitudinal impedance and phase angle as a function of frequency based on resonance and anti-resonance frequencies for [001]-oriented bar PZN-9PT crystals; (**b**) Thickness impedance and phase angle as a function of frequency for [001]-oriented square PZN-9PT crystals; (**c**,**d**) P–E loops of [001]-oriented PZN-9PT crystal at different temperatures and electric fields, respectively.

2.3. Nano Twin Domains of PZN-9PT Crystals

One probable explanation for the excellent properties of PZN–PT is the adaptive phase model based on nanoscale twinned O and T domains [24], which is evidenced in this work. The dark field TEM image (Figure 5a) shows a large number of domain-like bend striations in the order of several tens of angstrom units wide, indicating a typical relaxor domain structure. Similar domain striations were previously reported by Xu et al. [25] in the compositions near the MPB of PMN-xPT and were designated as tweed-like structures. The selected area electronic diffraction (SAED) pattern captured along the pseudocubic [001] zone axis (Figure 5b) shows that the 220 and 2-20 reflections are perpendicular to each other strictly, as illustrated by the blue dashed square. This result implies that the phase of as-grown PZN-9PT belongs neither to a monoclinic nor to a rhombohedral structure, well consistent with the XRD results. Additionally, the single crystals of PZN-9PT exhibit twin structures, as suggested by the red curve line in the upper left of Figure 5b from the profile line of the emission diffraction spot (as marked by the red dashed line in Figure 5b). This result can be further confirmed by the high resolution TEM (HRTEM) images, as shown in Figure 5d, corresponding to the inverse Fast Fourier Transformation (FFT) image from Figure 5c. The average domain size has been estimated to be about 8.3 nm in width, as labeled by the yellow double arrow in Figure 5d. Probably, the twin structures induce the crystal of PZN-9PT to exhibit the outstanding piezoelectric properties, as demonstrated in Figure 4. Moreover, from the inverse FFT image (Figure 5f) of the blue rectangle area in Figure 5e, we can calculate the lattice parameters of *a* and *b* for the PZN-9PT crystals to be about 4.05 Å and 4.03 Å, respectively, which are well consistent with the XRD results (O-phase: a = 4.031 Å, b = 4.05 Å; T-phase: a = b = 4.036 Å). Based on above observations, the correlation between the structure evolution and domain configuration reveals something interesting and important for understanding the origin of the giant piezoelectric response in PZN-PT. The continuous behavior of the phase evolution implies that PZN-9PT is in an instable regime, which may be the key factor responsible for its ultrahigh piezoelectric coefficient. Temperature-dependent TEM and/or in-situ synchrotron white-beam X-ray microdiffraction analyses [26] should be conducted in the future to establish possible correlations. It can be reliably speculated that the coexistence of the T-phase and O-phase results in large amounts of nanoscale twin domains, while the coexistence of the T-phase and the C-phase lead to nanoscale domains (including nano twin domains) and PNRs at around phase transition points and thus is responsible for the diffuse dielectric peaks.

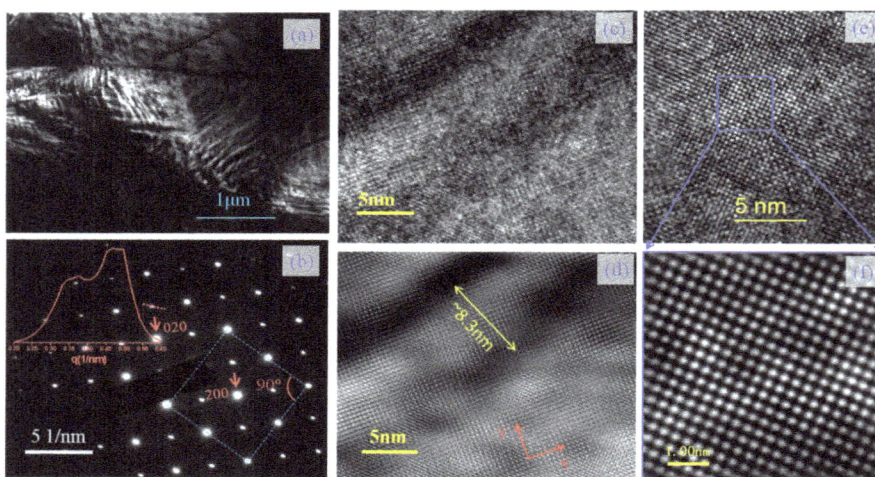

Figure 5. TEM images: (**a**) low TEM image; (**b**) SAED image; (**c**,**e**) high resolution TEM image; (**d**) corresponding to the inverse FFT image of (**c**); (**f**) corresponding to the inverse FFT image of the blue rectangle area in (**e**).

3. Materials and Methods

3.1. Fabrication of PZN-9PT Single Crystal

The PZN-9PT single crystals were grown with a top-seeded solution. The start materials, PbO (99.9%), TiO_2 (99.9%), ZnO (99.9%), and Nb_2O_5 (99.9%), were mixed according to the designed composition of PZN-9PT. Here, 10% of ZnO in excess of the stoichiometric ratio mole was added in the mixture to suppress the pyrochlore phase and decrease segregation simultaneously. The mixture of PbO and H_3BO_3 (99.9%) was used as a flux. The growing process was similar to the ones described in [27,28]. The weighed chemicals were thoroughly mixed and loaded into a platinum crucible with a size of $\Phi\, 40 \times 50\ mm^2$, which were then placed into a vertical tube furnace equipped with an automatic temperature controller to melt. The furnace was heated from room temperature to 1050 °C at a rate of 100 °C/h, held for 20 h, then slowly cooled from 1050 to 970 °C at a rate of 5 °C/h, and then from

940 to 900 °C at a rate of 0.5 °C/h. At the end of the slow cooling process, the as-grown crystal was pulled out of the melt and then annealed to room temperature at a rate of 20 °C/h.

3.2. Characterization Procedure

Firstly, the crystal component was analyzed by inductively coupled plasma optical emission spectrometry (ICP-OES). Then, in situ high-temperature XRD data were collected using the D8 Advance (Bruker, Karlsruhe, Germany) high temperature-diffractometer equipped with Cu Ka radiation and a graphite monochromator. The scan step is of 0.02° (2θ) with an angular range of 10°–70°. The as-grown single crystals were sliced into plates and bars along the [001] direction with different dimensions for the electric measurements. A sample with dimensions of $4^L \times 4^W \times 0.5^T$ mm^3 was used for dielectric and piezoelectric measurements, and another sample with dimensions of $3.5^L \times 2.5^W \times 0.6^T$ mm^3 for ferroelectric measurements. For resonance and anti-resonance measurements, a sample with dimensions of $0.9^L \times 0.9^W \times 3.12^T$ mm^3 was prepared. All the samples were polished and coated with silver paste as electrodes. The dielectric properties were measured using a computer-controlled Alpha-A broadband dielectric/impedance spectrometer (Novocontrol GmbH, Montabaur, Germany), with an AC signal of 0.3 V (peak-to-peak) applied. Measurements were carried out from −30 to 300 °C with a step of 2 °C. The same setup was used to measure the resonance frequency (f_r) and anti-resonance frequency (f_a) of a bar sample. Poling was performed by applying a DC electric field of 15 kV/cm along the [001] direction of the crystal at 120 °C for 15 min, and then by keeping the electric field on while cooling down to room temperature. The piezoelectric coefficient was measured using a quasi-static d_{33} meter (Institute of Acoustics, Chinese Academy of Sciences, model ZJ-4AN, Beijing, China). The polarization-electric field (*P–E*) hysteresis loop was tested using an aix-ACCT TF2000 analyzer. The microstructure of the crystal sample was studied by TEM (JEOL, JEM-2100F, Tokyo, Japan). The out-of-plane feature of the ferroelectric domains was investigated by a piezoelectric force microscope (Asylum Research, Oxford, UK).

4. Conclusions

To sum up, the evolution of phase transition of the relaxor ferroelectric PZN-9PT from the orthorhombic to cubic structure is a continuous process rather than a sharp process, as demonstrated by the variable XRD and Raman spectra. This phenomenon is very similar to the phase transition of paraffin from solid to liquid. Resonance and anti-resonance spectra show that the PZN-9PT single crystals exhibit excellent piezoelectric properties, which probably originates from their twin structures, as demonstrated by TEM.

Acknowledgments: This work was supported by the Chinese Postdoctoral Science Foundation (No. 2015M572356), the National Natural Science Foundation of China (Nos. 11604214, 11604140 and 21405106), the Hong Kong, Macao and Taiwan Science & Technology Cooperation Program of China (No. 2015DFH10200), and the Science and Technology Research Items of Shenzhen (No. JCYJ20160422102802301 & KQJSCX20160226195624).

Author Contributions: Hua Zhou and Shanming Ke conceived the idea and designed the experiments; Tao Li prepared the PZN-PT single crystals; Hua Zhou, Nian Zhang, Manfang Mai, Peng Lin and Chuanwei Huang performed the experiments; Xierong Zeng, Haitao Huang, Mao Ye and Shanming Ke analyzed the data and wrote the paper; Haitao Huang, Manfang Mai and Shanming Ke revised the paper. All authors discussed the results and have given approval to the final version of the manuscript.

Conflicts of Interest: The authors declare no conflict of interest.

References

1. Kuwata, J.; Uchino, K.; Nomura, S. Dielectric and piezoelectric properties of 0.91Pb(Zn$_{1/3}$Nb$_{2/3}$)O$_3$-0.09PbTiO$_3$ single crystals. *Jpn. J. Appl. Phys.* **1982**, *21*, 1298–1302. [CrossRef]

2. Zhang, S.J.; Li, F.; Jiang, X.N.; Kim, J.; Luo, J.; Geng, X.C. Advantages and challenges of relaxor-PbTiO$_3$ ferroelectric crystals for electroacoustic transducers—A review. *Prog. Mater. Sci.* **2015**, *68*, 1–66. [CrossRef] [PubMed]

3. Sun, E.W.; Cao, W.W. Relaxor-based ferroelectric single crystals: Growth, domain engineering, characterization and applications. *Prog. Mater. Sci.* **2014**, *65*, 124–210. [CrossRef] [PubMed]

4. Zhang, S.J.; Li, F.J. High performance ferroelectric relaxor-PbTiO$_3$ single crystals: Status and perspective. *J. Appl. Phys.* **2012**, *111*, 031301. [CrossRef]

5. Zhang, S.J.; Xia, R.; Lebrun, L.; Anderson, D.; Shrout, T.R. Piezoelectric materials for high power, high temperature applications. *Mater. Lett.* **2005**, *59*, 3471–3475. [CrossRef]

6. Li, T.; Li, X.Z.; Guo, D.; Wang, Z.J.; Liu, Y.; He, C.; Chu, T.; Ai, L.D.; Pang, D.F.; Long, X.F. Phase diagram and properties of high T_C/T_{R-T}Pb(In$_{1/2}$Nb$_{1/2}$)O$_3$-Pb(Zn$_{1/3}$Nb$_{2/3}$)O$_3$-PbTiO$_3$ ferroelectric ceramics. *J. Am. Ceram. Soc.* **2013**, *96*, 1546–1553. [CrossRef]

7. Haertling, G.H. Ferroelectric ceramics: history and technology. *J. Am. Ceram. Soc.* **1999**, *82*, 797–818. [CrossRef]

8. Samara, G.A. The relaxational properties of compositionally disordered ABO$_3$ perovskites. *J. Phys. Condens. Matter* **2003**, *15*, R367–R411. [CrossRef]

9. Ye, Z.-G. Crystal chemistry and domain structure of relaxor piezocrystals. *Curr. Opin. Solid State Mater.* **2002**, *6*, 35–44. [CrossRef]

10. Ye, Z.G. Relaxor ferroelectric complex pervoskites: Structure, properties and phase transitions. *Key Eng. Mater.* **1998**, *155*, 81–122. [CrossRef]

11. Janga, H.M.; Kim, S.C. Pb(B′$_{1/2}$B″$_{1/2}$)O$_3$-type perovskites: Part I. Pair-correlation theory of order-disorder phase transition. *J. Mater. Res.* **1997**, *12*, 2117–2126. [CrossRef]

12. Wang, Y.J.; Wang, D.; Yuan, G.L.; Ma, H.; Xu, F.; Li, J.F.; Viehland, D.; Gehring, P.M. Fragile morphotropic phase boundary and phase stability in the near-surface region of the relaxor ferroelectric (1−x)Pb(Zn$_{1/3}$Nb$_{2/3}$)O$_3$-xPbTiO$_3$: [001] Field-cooled phase diagrams. *Phys. Rev. B* **2016**, *94*, 174103. [CrossRef]

13. Cox, D.E.; Noheda, B.; Shirane, G.; Uesu, Y.; Fujishiro, K.; Yamada, Y. Universal phase diagram for high-piezoelectric perovskite systems. *Appl. Phys. Lett.* **2001**, *79*, 400–402. [CrossRef]

14. He, C.J.; Xu, F.; Wang, J.M.; Liu, Y.W. Refractive index dispersion of relaxor ferroelectric 0.9Pb(Zn$_{1/3}$Nb$_{2/3}$)O$_3$-0.1PbTiO$_3$ single crystal. *Cryst. Res. Technol.* **2009**, *44*, 211–214. [CrossRef]

15. Slodczyk, A.; Colomban, P. Probing the nanodomain origin and phase transition mechanisms in (un)poled PMN-PT single crystals and textured ceramics. *Materials* **2010**, *3*, 5007–5028. [CrossRef]

16. Ye, Z.-G.; Dong, M.; Zhang, L. Domain structures and phase transitions of the relaxor-based piezo-/ferroelectric (1−x)Pb(Zn$_{1/3}$Nb$_{2/3}$)O$_3$-xPbTiO$_3$ single crystals. *Ferroelectrics* **1999**, *229*, 223–232. [CrossRef]

17. Chang, W.S.; Lim, L.C.; Yang, P.; Ku, C.-S.; Lee, H.-Y.; Tu, C.-S. Transformation stress induced metastable tetragonal phase in (93–92)%Pb(Zn$_{1/3}$Nb$_{2/3}$)O$_3$-(7–8)%PbTiO$_3$ single crystals. *J. Appl. Phys.* **2010**, *108*, 044105. [CrossRef]

18. Uesu, Y.; Matsuda, M.; Yamada, Y.; Fujishiro, K.; COX, D.E.; Noheda, B.; Shirane, G. Symmetry of high-piezoelectric Pb-based complex perovskites at the morphotropic phase boundary: I. neutron diffraction study on Pb(Zn$_{1/3}$Nb$_{2/3}$)O$_3$-9%PbTiO$_3$. *J. Phys. Soc. Jpn.* **2002**, *71*, 960–965. [CrossRef]

19. Mishra, K.K.; Arora, A.K.; Tripathy, S.N.; Pradhan, D. Dielectric and polarized Raman spectroscopic studies on 0.85Pb(Zn$_{1/3}$Nb$_{2/3}$)O$_3$-0.15PbTiO$_3$ single crystal. *J. Appl. Phys.* **2012**, *112*, 073521. [CrossRef]

20. Cheng, J.; Yang, Y.; Tong, Y.H.; Lu, S.B.; Sun, J.Y.; Zhu, K.; Liu, Y.L.; Siu, G.G.; Xu, Z.K. Study of monoclinic-tetragonal-cubic phase transition in Pb(Zn$_{1/3}$Nb$_{2/3}$)O$_3$-0.08PbTiO$_3$ single crystals by micro-Raman spectroscopy. *J. Appl. Phys.* **2009**, *105*, 053519. [CrossRef]

21. *IEEE Standard on Piezoelectricity: An American National Standard*; Institute of Electricity and Electronics Engineers: New York, NY, USA, 1979.

22. Zhang, S.J.; Randall, C.A.; Shrout, T.R. Characterization of perovskite piezoelectric single crystals of 0.43BiScO$_3$-0.57PbTiO$_3$ with high Curie temperature. *J. Appl. Phys.* **2004**, *95*, 4291–4295. [CrossRef]

23. Harada, K.; Hosono, Y.; Saitoh, S.; Yamashita, Y. Crystal growth of Pb[(Zn$_{1/3}$Nb$_{2/3}$)$_{0.91}$Ti$_{0.09}$]O$_3$ using a crucible by the supported Bridgman method. *Jpn. J. Appl. Phys.* **2000**, *33*, 3117–3120. [CrossRef]

24. Jin, Y.M.; Wang, Y.U.; Khachaturyan, A.G.; Li, J.F.; Viehland, D. Conformal miniaturization of domains with low domain-wall energy: Monoclinic ferroelectric states near the morphotropic phase boundaries. *Phys. Rev. Lett.* **2003**, *91*, 197601. [CrossRef] [PubMed]

25. Xu, Z.; Kim, M.; Li, J.F.; Viehland, D. Observation of a sequence of domain-like states with increasing disorder in ferroelectrics. *Philos. Mag. A* **1996**, *74*, 395–406. [CrossRef]

Crystals **2017**, *7*, 130

26. Li, T.; Du, Z.H.; Tamura, N.; Lu, W.; Ye, M.; Zeng, X.R.; Ke, S.M.; Huang, H.T. In-situ Synchrotron X-ray Micro-Beam Observation of Phase Transition and Nanotwin Domain Structure in (1–x)PZN-xPT Single Crystals. **2017**. in preparation.
27. Long, X.F.; Ling, J.B.; Li, X.Z.; Wang, Z.J.; Ye, Z.-G. Growth and Di-/Piezoelectric properties of Al-doped PMN-30PT single crystals. *Cryst. Growth Des.* **2009**, *9*, 657–659. [CrossRef]
28. Long, X.F.; Ye, Z.-G. Top-seeded solution growth and characterization of rhombohedral PMN-30PT piezoelectric single crystals. *Acta Mater.* **2007**, *55*, 6507–6512. [CrossRef]

© 2017 by the authors. Licensee MDPI, Basel, Switzerland. This article is an open access article distributed under the terms and conditions of the Creative Commons Attribution (CC BY) license (http://creativecommons.org/licenses/by/4.0/).

crystals

MDPI

Article

Phase Transition Behavior of the Layered Perovskite CsBi$_{0.6}$La$_{0.4}$Nb$_2$O$_7$: A Hybrid Improper Ferroelectric

Charlotte A. L. Dixon [1], Jason A. McNulty [1], Kevin S. Knight [2,3], Alexandra S. Gibbs [4] and Philip Lightfoot [1,*]

[1] School of Chemistry and EaStCHEM, University of St Andrews, St Andrews KY16 9ST, UK; cald@st-andrews.ac.uk (C.A.L.D.); jam242@st-andrews.ac.uk (J.A.M.)
[2] Department of Earth Sciences, University College London, Gower Street, London WC1E 6BT, UK; kevinstevenknight@gmail.com
[3] Department of Earth Sciences, The Natural History Museum, Cromwell Road, London SW7 5BD, UK
[4] ISIS Facility, Rutherford Appleton Laboratory, Chilton, Oxon OX11 0QX, UK; alexandra.gibbs@stfc.ac.uk
* Correspondence: pl@st-andrews.ac.uk; Tel.: +44-1334-463841

Academic Editor: Stevin Snellius Pramana
Received: 20 April 2017; Accepted: 10 May 2017; Published: 13 May 2017

Abstract: The phase behavior of the layered perovskite CsBi$_{0.6}$La$_{0.4}$Nb$_2$O$_7$, of the Dion-Jacobson family, has been studied by high-resolution powder neutron diffraction between the temperatures of 25 < T < 850 °C. At ambient temperature, this material adopts the polar space group $P2_1am$; this represents an example of hybrid improper ferroelectricity caused by the interaction of two distinct octahedral tilt modes. Within the limits of our data resolution, the thermal evolution of the crystal structure is consistent with a first-order transition between 700 and 750 °C, with both tilt modes vanishing simultaneously, leading to the aristotype space group $P4/mmm$. This apparent "avalanche transition" behavior resembles that seen in the related Aurivillius phase SrBi$_2$Nb$_2$O$_9$.

Keywords: polar crystal; layered perovskite; hybrid improper ferroelectric; phase transition

1. Introduction

Recent theoretical work has revealed novel mechanisms for designing non-centrosymmetric, polar and potentially ferroelectric materials, based on the coupling of two or more distinct structural distortions ("modes") [1–4]. So-called "hybrid improper ferroelectrics" (HIFs) have now been postulated for several families of layered perovskites, by the coupling of two distinct "octahedral tilt" modes with a polar mode. Although octahedral tilting has been a well-established phenomenon in perovskite crystallography for many years [5–7], in simple three-dimensional ABX$_3$ perovskites all the simple octahedral tilt schemes (as classified by Howard and Stokes [7]) give rise to strictly centrosymmetric crystals. It is now recognized that in more complex cases, for example cation-ordered or layered perovskites, inversion symmetry may be broken directly by particular combinations of octahedral tilt modes. The concept of HIFs has been typically applied by theorists to mean the imposition of a stable polar ground state in a crystal via the coupling of two or more intrinsically non-polar modes with a polar mode ("tri-linear coupling"), with no requirement for an experimentally-realized polarization reversal. However, such experimental demonstrations of polarization-field hysteresis have now been reported, for example in the Ruddlesden-Popper (RP) phase Ca$_3$Ti$_2$O$_7$ [8] and the Dion-Jacobson (DJ) phase RbBiNb$_2$O$_7$ [9].

Prior to the recent flurry of work from the theoretical standpoint, it was known that ferroelectricity, or at least polarity, co-existed with octahedral tilting in several layered perovskites, the most well-established being the Aurivillius family, exemplified by SrBi$_2$Ta$_2$O$_9$ and Bi$_4$Ti$_3$O$_{12}$ [10]. Snedden et al. [11] later made the intuitive link between the nature of octahedral tilting in the DJ phase CsBiNb$_2$O$_7$ and that in SrBi$_2$Ta$_2$O$_9$; both have a tilt system analogous to that described by the Glazer notation a$^-$a$^-$c$^+$ in ABX$_3$ perovskites. This leads to related superlattices and space groups in both SrBi$_2$Ta$_2$O$_9$ and CsBiNb$_2$O$_7$, viz. superlattices of ~$\sqrt{2}a_t$ x $\sqrt{2}a_t$ x c_t relative to the aristotype tetragonal phase in each case, and polar space groups $A2_1am$ and $P2_1am$ in SrBi$_2$Ta$_2$O$_9$ and CsBiNb$_2$O$_7$, respectively. The nature of these superlattices and the allowed space groups of the possible distorted structures may be systematized by considering which particular distortion modes give rise to the specific tilts within each family of layered perovskites. So, for example, the "out-of-phase tilt", a$^-$a$^-$c^0, and the "in-phase tilt", a^{0}a^{0}c$^+$, correspond to the modes with irreducible representations M$_5$$^-$ and M$_2$$^+$, respectively, for the DJ phases of the type AA'B$_2$O$_7$, for which the parent phase is described in space group $P4/mmm$ (these are "$n = 2$" examples of the generic DJ composition AA'$_{n-1}$B$_n$O$_{3n+1}$, with n representing the number of octahedral layers per perovskite-like block). The coupling of the two modes leads naturally to the polar space group $P2_1am$ for CsBiNb$_2$O$_7$ and, in principle, permits ferroelectricity. These ideas have been described in more detail by Benedek et al. [3,4].

Although the a$^-$a$^-$c$^+$ tilt system is predicted to be a stable option for DJ phases of the type AA'B$_2$O$_7$, recent experimental work has shown that other related polymorphs are possible, and a tilt system a$^-$b^{0}c$^+$, in polar space group $Amm2$, has been reported in CsLaNb$_2$O$_7$ in the range $350 < T < 550$ K [12] (curiously an earlier single crystal X-ray study of CsLaNb$_2$O$_7$ reported the adoption of the aristotype tetragonal structure, $P4/mmm$ at 296 K [13]). Moreover, a recent experimental and theoretical study of the triple-layer DJ phase CsBi$_2$Ti$_2$NbO$_{10}$ has suggested that a proper rather than improper mechanism drives the polarity of the ambient temperature phase [14]. It is therefore of interest to explore more fully the phase behavior of further examples of the DJ family in order to map out the occurrence of the possible tilt systems, especially those giving rise to polarity, and also the nature of phase transitions in these systems as a function of T.

In our earlier work [11], it was established that both CsBiNb$_2$O$_7$ and CsNdNb$_2$O$_7$ adopted the polar $P2_1am$ space group at ambient temperature, with the latter composition displaying a much reduced orthorhombic distortion relative to the former. Later, Fennie and Rabe predicted ferroelectricity in CsBiNb$_2$O$_7$ from a first-principles DFT study [15]. Our follow-up experimental studies [16] on CsBiNb$_2$O$_7$ suggested that ferroelectricity was hindered by moisture uptake, leading to proton conductivity and leakage. Nevertheless, a more recent work by Chen et al. [17] did provide experimental evidence for ferroelectricity in this compound. Goff et al. [16], attempted to determine any high temperature phase transitions in CsBiNb$_2$O$_7$ using powder neutron diffraction: no evidence for phase transitions was identified up to 900 °C, above which sample deterioration precluded further study. By analogy with CsNdNb$_2$O$_7$, we speculated that La-doping of CsBiNb$_2$O$_7$ would reduce the orthorhombic distortion such that any phase transition to higher symmetry might be brought down in temperature relative to that of CsBiNb$_2$O$_7$, and make this transition more amenable to experimental study.

In this paper, we report the phase behavior of CsBi$_{0.6}$La$_{0.4}$Nb$_2$O$_7$; this composition was chosen since CsBiNb$_2$O$_7$ and CsLaNb$_2$O$_7$ have been shown to adopt different phases at ambient temperature, and it was anticipated that a phase transition(s) might be observable within a reasonable temperature regime.

2. Results

Ambient temperature structure: Preliminary studies focused on the solid solution CsBi$_{1-x}$La$_x$Nb$_2$O$_7$ (see Supplementary Information), which is confirmed to exist for all compositions, $0 < x < 1$. The $x = 0.4$ composition, i.e., CsBi$_{0.6}$La$_{0.4}$Nb$_2$O$_7$, was chosen for further study by high-resolution powder neutron diffraction. Initial analysis of the room-temperature powder neutron diffraction (PND) data assumed the $P2_1am$ model previously identified for CsBiNb$_2$O$_7$ [11]. Rietveld refinement suggested this model to be correct, with an excellent fit, accounting for all observed peak splittings and superlattice peaks (Figure 1). However, in order for this assignment to be unambiguous, other options need to be considered. There are two aspects to this: first, the unit cell metrics need to be determined and second, the nature of any characteristic superlattice peaks (especially those arising from octahedral tilting) needs to be established.

Considering the unit cell metrics first: the possible superlattice options to be considered are either an approximate doubling or quadrupling of the unit cell in the *ab* plane; i.e., $a \sim b \sim \sqrt{2}a_t$ or $a \sim b \sim 2a_t$. Superlattices along the *c*-axis are not considered in the simplest scheme (and are not necessary in this case). These two options can easily be distinguished based on related peak splittings observed in the experimental PND data: the $\sqrt{2}a_t$ metric splits peaks of the type (*hhl*) in the parent tetragonal phase, whereas the $2a_t$ metric splits peaks of the type (*h0l*/*0kl*). Specific peaks labelled in Figure 1 confirm unambiguously that an orthorhombic (or lower) symmetry unit cell of metrics $a \sim b \sim \sqrt{2}a_t$, $c \sim c_t$, is required to account for all the observed peak splittings.

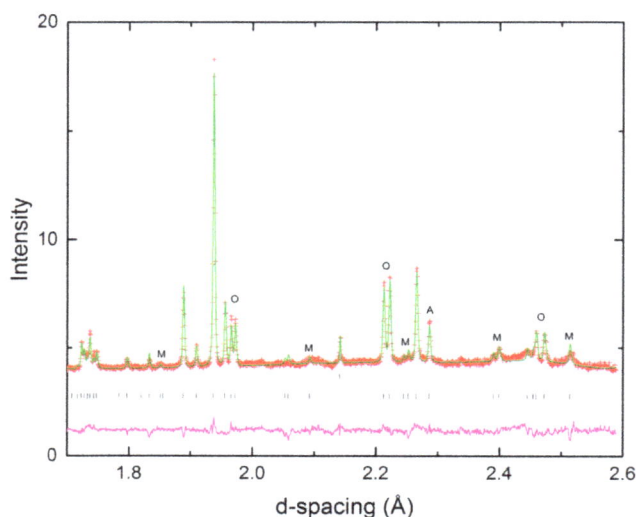

Figure 1. Portion of the Rietveld plot for the $P2_1am$ model at ambient temperature. Peaks marked 'M' represent superlattice peaks at the *M*-point; peaks marked "O" are subcell peaks that split into doublets in this superlattice ($a \sim b \sim \sqrt{2}a_t$), whereas the peak marked "A" represents a peak that would be split if the alternative supercell ($a \sim b \sim 2a_t$) was present. The peak at $d \sim 2.14$ Å is from the vanadium sample holder.

Strayer et al. [12] listed all models corresponding to the simple combinations of octahedral tilts for this structural family (see Supplementary information in Reference [12]). Specifically, the definition of simple combinations here means that there are four distinct octahedral tilts modes to be considered, viz. in-phase and out-of-phase tilts of two adjacent octahedra along the *a* (and/or *b*) and *c* axes of the parent tetragonal unit cell. These four modes are given the Glazer-like symbols a^+ (or b^+) and c^+ for the respective in-phase tilts and a^- (b^-) and c^- for the out-of-phase tilts. In turn, these distortions give rise to superlattice peaks in the diffraction pattern at characteristic points, labelled M ($k = \frac{1}{2}, \frac{1}{2}, 0$) or X ($k = 0, \frac{1}{2}, 0$) relative to the parent tetragonal reciprocal lattice. The specific modes corresponding to each of the four possible tilts are X_1^+ (a^+), M_5^- (a^-) M_2^+ (c^+) and M_4^- (c^-). The nature of these modes and their effect on the diffraction pattern can be probed using the on-line software ISODISTORT [18]. The *X*-point mode can be easily ruled out (no peaks are visible at the relevant positions), but the *M*-point modes provide some ambiguity, as they may contribute to many of the same superlattice peaks. In fact, closer inspection of the allowed superlattice intensities from each of the *M*-point modes reveals that the M_4^- mode can also be ruled out: we also note that neither the X_1^+ nor the M_4^- modes were considered in the earlier theoretical studies [3,15]. The M_5^- mode is certainly required, as it is the only contributor to certain peaks (for example the one near d ~2.5 Å). The M_2^+ mode contributes to several observed peaks (for example those near d ~2.4 Å), but the M_5^- mode may also contribute to those. Table 1 lists an abbreviated selection of the possible octahedral-rotation-induced superlattices of Strayer et al. Taking into account the requirement of a $\sqrt{2}a_t \times \sqrt{2}a_t \times c_t$ orthorhombic metric and the presence of at least the M_5^- mode (and also possibly the M_2^+ mode) reveals the only plausible models are the $P2_1am$ and $Pmam$ ones, as suggested by Snedden et al. [11] for $CsBiNb_2O_7$. In particular, the $Amm2$ model observed for $CsLaNb_2O_7$ can be ruled out in the present case. In order to confirm $P2_1am$ as the correct model for $CsBi_{0.6}La_{0.4}Nb_2O_7$ at ambient temperature, a comparative refinement against the $Pmam$ model was carried out. This led to agreement factors of χ^2 ~5.870 and 9.081, for the $P2_1am$ and $Pmam$ models. $P2_1am$ is therefore selected as the best model at ambient T. The corresponding crystal structure is shown in Figure 2, refined atomic parameters are given in Table 2 and selected bond lengths in Table 3.

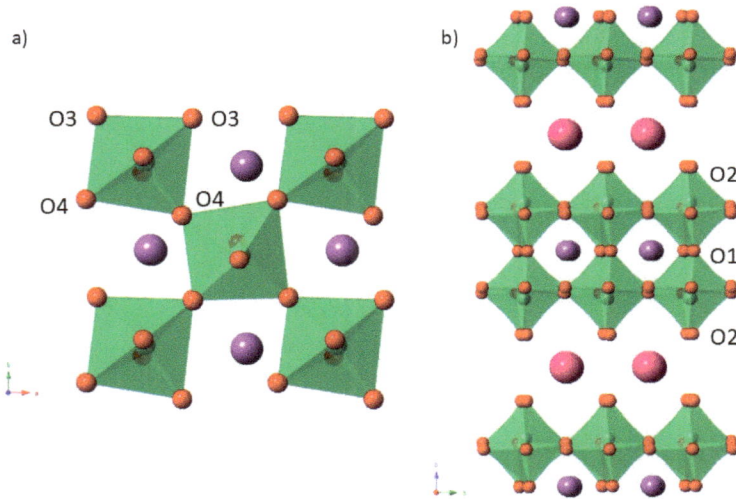

Figure 2. Crystal structure of $CsBi_{0.6}La_{0.4}Nb_2O_7$ at ambient temperature (**a**) down the c-axis, showing the in-phase (M_2^+) tilt and (**b**) perpendicular view, showing the out-of-phase (M_5^-) tilt.

Table 1. Example models based on combinations of octahedral tilt modes (adapted from Strayer et al. [12].)

Glazer Notation	Active Modes	Space Group	Metrics
$a^0a^0c^0$	None	$P4/mmm$	$a_t \times a_t \times c_t$
$a^0a^0c^+$	M_2^+	$P4/mbm$	$\sqrt{2}a_t \times \sqrt{2}a_t \times c_t$
$a^0a^0c^-$	M_4^-	$P4/nbm$	$\sqrt{2}a_t \times \sqrt{2}a_t \times c_t$
$a^+a^+c^0$	X_1^+	$P4/mmm$	$2a_t \times 2a_t \times c_t$
$a^-a^-c^0$	M_5^-	$Pmam$	$\sqrt{2}a_t \times \sqrt{2}a_t \times c_t$
$a^-a^-c^+$	$M_2^+ \otimes M_5^-$	$P2_1am$	$\sqrt{2}a_t \times \sqrt{2}a_t \times c_t$
$a^-b^0c^0$	M_5^-	$Cmmm$	$2a_t \times 2a_t \times c_t$
$a^-b^0c^+$	$M_2^+ \otimes M_5^-$	$Amm2$ *	$c_t \times 2a_t \times 2a_t$
$a^-b^-c^0$	M_5^-	$P2/m$	$\sqrt{2}a_t \times c_t \times \sqrt{2}a_t$
$a^-b^-c^+$	$M_2^+ \otimes M_5^-$	Pm	$\sqrt{2}a_t \times c_t \times \sqrt{2}a_t$

* $C2mm$ setting for c corresponding to c_t.

Table 2. Refined structural model for $CsBi_{0.6}La_{0.4}Nb_2O_7$ at 25 °C. Space group $P2_1am$, a = 5.49618 (14), b = 5.46012 (14), c = 11.3254 (3) Å.

Atom	Wyckoff Position	x	y	z	$100 * U_{iso}$ ($Å^2$)
Cs1	2b	0.3349 (17)	0.2579 (12)	0.5	2.08 (10)
Bi/La1 *	2a	0.367	0.2692 (8)	0.0	2.91 (11)
Nb1	4c	0.3334 (10)	0.7540 (6)	0.20533 (14)	0.56 (7)
O1	2a	0.321 (2)	0.6892 (9)	0.0	2.83 (15)
O2	4c	0.3272 (12)	0.7812 (7)	0.3574 (2)	1.66 (9)
O3	4c	0.0869 (15)	0.0104 (12)	0.1589 (3)	3.10 (13)
O4	4c	0.5290 (11)	0.4590 (8)	0.1840 (3)	2.25 (13)

* fixed occupancy $Bi_{0.6}La_{0.4}$; the x-coordinate is fixed to define the origin of the polar axis.

Table 3. Selected bond lengths (Å) for $CsBi_{0.6}La_{0.4}Nb_2O_7$ at 25 °C in the $P2_1am$ model.

Cs-O	Bond Length (Å)	Bi/La-O	Bond Length (Å)	Nb-O	Bond Length (Å)
Cs1-O2 × 2	3.063 (7)	Bi1-O1	2.307 (7)	Nb1-O1	2.3531 (19)
Cs1-O2 × 2	3.158 (10)	Bi1-O1	2.508 (12)	Nb1-O2	1.730 (3)
Cs1-O2 × 2	3.231 (10)	Bi1-O3 × 2	2.651 (6)	Nb1-O3	1.968 (9)
Cs1-O2 × 2	3.282 (7)	Bi1-O3 × 2	2.758 (6)	Nb1-O3	2.018 (8)
		Bi1-O4 × 2	2.492 (4)	Nb1-O4	1.952 (6)
				Nb1-O4	2.052 (6)

Structural Evolution Versus *T*: The continued presence of orthorhombic distortion can easily be determined on raising the temperature, by tracking the O peak splittings highlighted in Figure 1. However, since the *M*-point superlattice peaks are very weak, even at ambient temperature, it is problematic to use a simple qualitative visual observation of these as a single reliable measure of the structural distortions versus *T*. Instead, full comparative Rietveld refinement of the structure in both $P2_1am$ and $Pmam$ models is the preferred method of discrimination (since the tilts significantly affect the intensities of the subcell peaks, in addition to the supercell peaks). Both models were therefore refined against each data set up to 725 °C, taking care to use a consistent set of parameters throughout (see Experimental). At 700 °C the orthorhombic splitting is still clearly seen in the raw data, but at 725 °C the splitting disappears, and the unit cell becomes metrically tetragonal (Figure 3). The parent $P4/mmm$ model was therefore trialed at 725 °C and above, with a further check on the alternative possible tetragonal model, $P4/mbm$ ($a^0a^0c^+$, Table 1) at 725 °C only. In addition to tracking the comparative fit quality for the two orthorhombic models, ISODISTORT was used to derive the mode amplitudes versus *T*. Thermal evolution of the unit cell parameters is given in Figure 4, and for the distortion modes in Figure 5. Comparative goodness-of-fit parameters (χ^2) for the $P2_1am$ and $Pmam$ models are given in Figure 6.

The difference in χ^2 values clearly supports the assignment of the $P2_1am$ rather than the $Pmam$ model up to 700 °C (χ^2 values are 2.055 and 2.273 for $P2_1am$ and $Pmam$, respectively, at 700 °C). In addition, the tilt mode amplitudes (Figure 5) tend towards zero, but do not become zero, at 700 °C. The negligible difference in fit quality between the four models tested at 725 °C (χ^2 values 2.065, 2.054, 2.041 and 1.883 for $P4/mmm$, $P4/mbm$, $Pmam$ and $P2_1am$, respectively) supports the absence of both tilt modes at this temperature (we note that the refinement in the $P2_1am$ model was unstable to the refinement of thermal parameters, thus a direct comparison is not possible). Although the unit cell volume has no clear discontinuity within the temperature resolution of our datasets, the fact that the polar mode $\Gamma_5{}^-$ seems to have a step-like change at the transition may indicate the existence of a first-order component.

We attempted to fit the thermal evolution of the tilt modes, $M_2{}^+$ and $M_5{}^-$, to critical behavior of the form: amplitude = A $(T_c - T)^\beta$. This produced the fit parameters of A = 0.0292, 0.227; T_C = 749, 742 °C; β = 0.401, 0.127 for the $M_2{}^+$ and $M_5{}^-$ modes, respectively. The key observation here is that the derived value of T_C is approximately equal for the two modes, implying that they condense simultaneously. Unfortunately, the paucity of data in the vicinity of T_C prohibits a definitive conclusion about the exact

T_C and β values, and also the specific mode which drives the transition. However, further support for our crystallographic model comes from a measurement of dielectric properties versus T. Figure 7 shows a single dielectric maximum just below 750 °C, with no evidence of any lower temperature anomalies. This supports an abrupt, single phase transition, as was observed crystallographically.

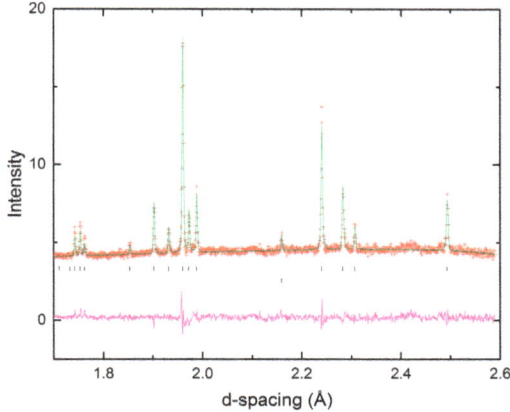

Figure 3. Portion of the Rietveld plot for the $P4/mmm$ model at 725 °C. Note the absence of both M superlattice peaks and O splittings, compared to Figure 1.

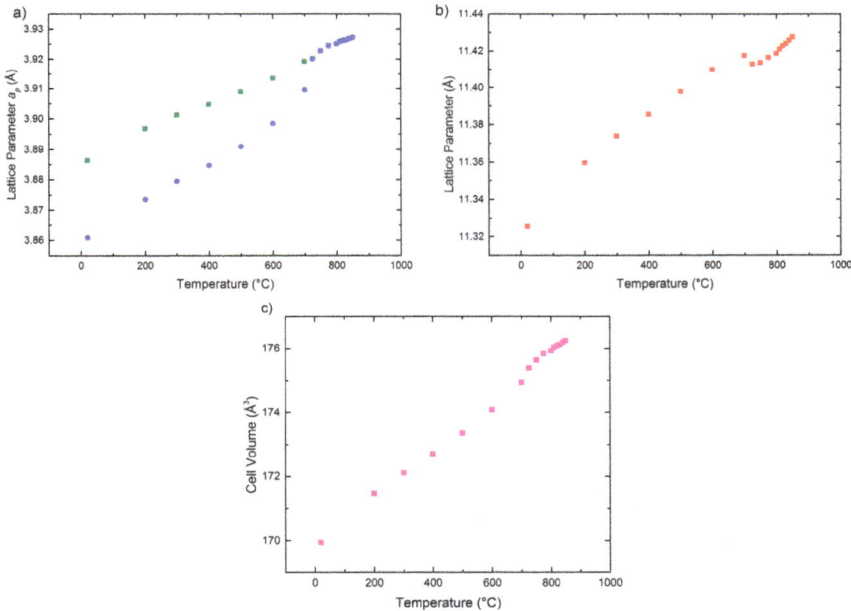

Figure 4. Thermal evolution of the unit cell parameters. (**a**) *a* (squares) and *b* (circles) parameters; (**b**) *c* parameter; (**c**) unit cell volume, suggesting a first-order orthorhombic-tetragonal phase transition near 725 °C. Note that the *a* and *b* parameters in the orthorhombic phase are normalized to the tetragonal subcell (i.e., divided by $\sqrt{2}$).

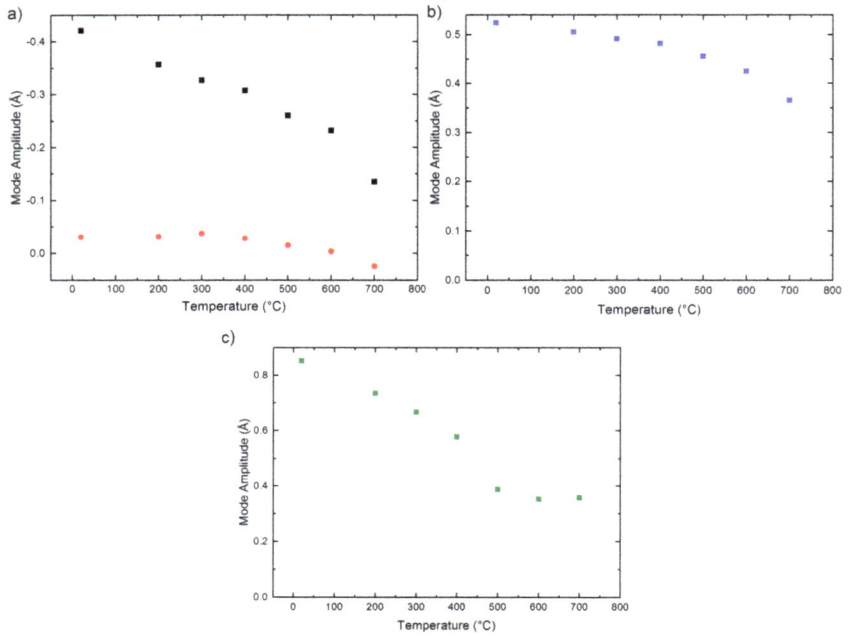

Figure 5. Thermal evolution of the mode amplitudes (from ISODISTORT). (**a**) M_2^+ (squares) and M_3^+ (circles) modes: M_2^+ is the in-phase tilt around *c*, M_3^+ is a minor octahedral distortion mode; (**b**) M_5^- mode: this mode is largely composed of contributions from the out-of-phase tilt around the *ab* plane, but also incorporates minor contributions from antiferrodistortive cation displacements along *b*; (**c**) Γ_5^- mode: this is the polar mode (i.e., polar displacive contributions from all atoms along the *a*-axis).

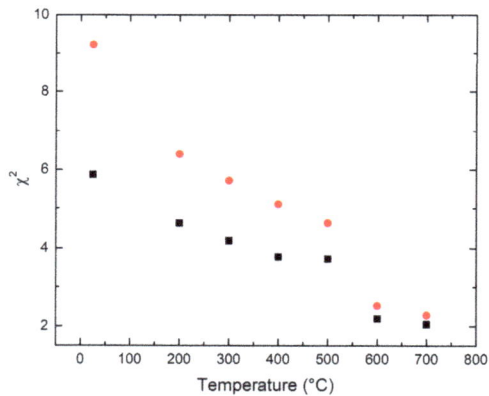

Figure 6. Comparison of Rietveld goodness-of-fit parameters (χ^2) for the $P2_1am$ and $Pmam$ models up to a temperature of 700 °C.

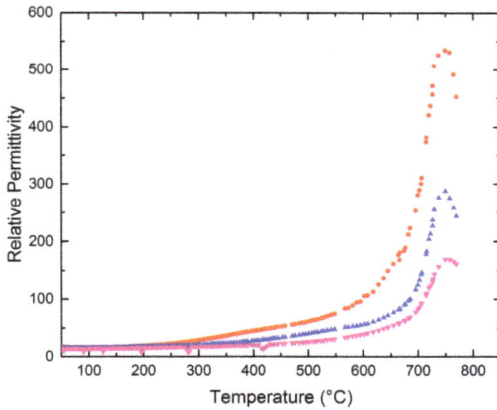

Figure 7. Relative permittivity data for $CsBi_{0.6}La_{0.4}Nb_2O_7$ at selected frequencies, obtained on cooling. Data collected at a frequency of 10 kHz are shown by red circles, 100 kHz by blue triangles and 1 MHz by pink triangles.

Taken together, these observations suggest a direct to transition from $P2_1am$ to $P4/mmm$ in the vicinity of 700–750 °C; we note that Landau theory requires the direct $P2_1am$ to $P4/mmm$ transition to be first-order, whereas the $Pmam$ to $P4/mmm$ transition is allowed to be continuous. Hence, the most likely scenario, which is also compatible with the present data, is that both tilt modes are lost simultaneously, with no substantial evidence for an intermediate phase of symmetry $Pmam$ or $P4/mbm$ (Table 1). The final refined model in $P4/mmm$ at 750 °C is given in Tables 4 and 5.

Table 4. Refined structural model for $CsBi_{0.6}La_{0.4}Nb_2O_7$ at 750 °C. Space group $P4/mmm$, a = 3.92282 (16), c = 11.4138 (5) Å.

Atom	Wyckoff Position	x	y	z	$100 * U_{iso}$ (Å^2)
Cs1	1b	0.0	0.0	0.5	7.09 (13)
Bi/La1 *	1a	0.0	0.0	0.0	7.17 (16)
Nb1	2h	0.5	0.5	0.20414 (16)	2.13 (9)
O1	4i	0.0	0.5	0.17054 (17)	5.56 (10)
O2	2h	0.5	0.5	0.3555 (3)	6.44 (13)
O3	1c	0.5	0.5	0.0	7.20 (17)

* fixed occupancy $Bi_{0.6}La_{0.4}$.

Table 5. Selected bond lengths (Å) for $CsBi_{0.6}La_{0.4}Nb_2O_7$ at 750 °C in the $P4/mmm$ model.

Cs-O	Bond Length (Å)	Bi/La-O	Bond Length (Å)	Nb-O	Bond Length (Å)
Cs1-O2 × 8	3.2273 (14)	Bi1-O1 × 8	2.7634 (14)	Nb1-O1 × 4	1.9985 (5)
		Bi1-O3 × 4	2.77385 (11)	Nb1-O2	1.727 (4)
				Nb1-O3	2.3300 (18)

3. Discussion and Conclusions

The present study indicates, within the temperature resolution of our data, a direct first-order transition from the parent, untilted phase, $P4/mmm$, to the observed ambient temperature phase, $P2_1am$ (tilt system $a^-a^-c^+$). This requires the simultaneous condensation of the two tilt modes M_5^- ($a^-a^-c^0$) and M_2^+ ($a^0a^0c^+$), which couple with the polar mode, Γ_5^-. This special case of tri-linear coupling has been termed an "avalanche transition" [19]. Due to the temperature increments used in this experiment, we cannot fully rule out a possible intermediate phase of symmetry *Pmam* or *P4/mbm* but, if present, this must only exist in a narrow temperature interval around 700 to 750 °C. Avalanche transitions are rare, but such a transition has been confirmed in the Aurivillius phase ferroelectric, $SrBi_2Nb_2O_9$ [20,21], which also displays the tilt system $a^-a^-c^+$. It has been suggested that $SrBi_2Nb_2O_9$ differs from its analogue $SrBi_2Ta_2O_9$, which does display an intermediate phase (tilt system $a^-a^-c^0$), because the magnitudes of the three necessary distortion modes in the ground state (i.e., ambient temperature) polar phase are relatively similar, rather than one tilt mode being dominant [21]. In their detailed study of $CsLaNb_2O_7$, Strayer et al. [12], reported a tilt system $a^-b^0c^+$ (space group *Amm*2, Table 1) between 550 and 350 K. This phase also requires the condensation of both the M_2^+ and M_5^- tilt modes, but differs from the present $P2_1am$ phase in the direction of the M_5^- mode. It might also be expected that an intermediate phase would be observed in the $P4/mmm$ to *Amm*2 pathway seen in $CsLaNb_2O_7$; i.e., via the sequential loss of either M_5^- then M_2^+, or vice versa. The corresponding intermediate phases (Table 1) in that case would be $a^0a^0c^+$ (*P4/mbm*) or $a^-b^0c^0$ (*Cmmm*). Strayer et al. ruled out these pathways by the observation of a clear SHG signal, signifying non-centrosymmetricity, and thus implying an avalanche transition in this composition also. As the present study represents the only detailed crystallographic study so far of the tilt transitions concerning a verified $P2_1am$ phase ($a^-a^-c^+$) in the n = 2 DJ family, it remains to be seen whether such avalanche transitions are common in the DJ family.

4. Experimental Section

4.1. Synthesis

A phase pure sample of $CsBi_{0.6}La_{0.4}Nb_2O_7$ was prepared using traditional ceramic methods. Stoichiometric amounts of La_2O_3 (99.9% Sigma-Aldrich, Dorset, UK), Nb_2O_5 (99.9% Alfa Aesar, Lancashire, UK) and a 20% excess of Cs_2CO_3 (99% Alfa Aesar, Lancashire, UK) were dried at 100 °C for 24 h. The loose powders were ground for a period of 30 min and pressed into pellets of approximately 10 mm diameter and 5 mm thickness. The pellets were annealed at 1000 °C for a period of 24 h with a cooling rate of 10 min^{-1}.

4.2. Powder Diffraction

Sample purity was gauged by preliminary X-ray diffraction using a PANalytical Empyrean (Cu $K_{\alpha1}$ radiation).

Neutron powder diffraction (NPD) was performed at beamline HRPD (High resolution powder diffractometer), ISIS facility, Oxfordshire, UK. A sample of approximately 5 g was placed in an 8-mm cylindrical vanadium can before loading into the diffractometer vacuum furnace. Patterns were collected at room temperature, and then at 100 intervals between 200–800 °C with a final data collection at 850 °C, each data collection lasting approximately 3.5 h, with the exception of the data collections at 600 and 700 °C (which lasted approximately 2 h). Intermediate patterns were collected at intervals of 10 between 810 and 840 °C, and 25 intervals between 725 and 775 °C, each data collection lasting approximately 25 min.

Rietveld refinement was performed using GSAS [22] and the EXPGUI interface [23]. The same set of standard parameters were used, including 3, 3, 21 and 5 parameters to model instrumental variables, scale factors, background and peak-shape for each dataset in the region of RT to 700 °C. The same refinement strategy was employed for the remaining high temperature datasets, except that the number of background terms was decreased from 21 to 18. A significant contribution from the vanadium sample can be seen throughout and this was fitted as a secondary phase at each temperature.

4.3. Dielectric Measurements

Pellets were electroded with sputtered Au before being coated with Ag paste (RS components). Dielectric measurements were made using a Wayne Kerr 6500B impedance analyzer with the sample mounted in a tube furnace. Capacitance and loss data were recorded in the frequency range of 100–10 MHz at a heating and cooling at a rate of 2 K min^{-1} over the temperature range of 50 to approximately 750 °C. Data collected on the cooling cycle are described here, since the heating cycle showed an additional anomaly due to the failure of the Ag paste (which is used to protect the Au electrode). Nevertheless, there was no evidence for hysteresis during the heating/cooling cycle.

Supplementary Materials: The following are available online at http://www.mdpi.com/2073-4352/7/5/135/s1, Figure S1: (**a**) normalised *a* (black squares) and *b* (red circles) lattice parameters obtained for varying values of *x* across the solid solution $CsBi_{1-x}La_xNb_2O_7$, (**b**) *c* lattice parameter for varying values of *x* and (**c**) (normalized) unit cell volume for varying values of *x*.

Acknowledgments: We thank EPSRC for a Ph.D. studentship to CALD (EP/P505097/1). We thank undergraduate students William Skinner and Adam Smyth for some preliminary work on the $CsBi_{1-x}La_xNb_2O_7$ solid solution and Finlay D. Morrison for guidance on the dielectric measurements.

Author Contributions: Charlotte A. L. Dixon carried out the synthesis and diffraction measurements. Kevin S. Knight and Alexandra S. Gibbs assisted with collection of neutron diffraction data and Jason A. McNulty carried out the dielectric measurements. Charlotte A. L. Dixon carried out the crystallographic analysis, with the assistance of Philip Lightfoot. Philip Lightfoot coordinated the project and wrote the paper, with the approval of all authors.

Conflicts of Interest: The authors declare no conflict of interest.

Appendix A

Further details of the crystal structures at selected temperatures may be obtained from Fachinformationszentrum (FIZ) Karlsruhe, 76344 Eggenstein-Leopoldshafen, Germany (e-mail: crysdata@fiz-karlsruhe.de) on quoting deposition numbers 432943-432945. The research data (raw neutron diffraction data) pertaining to this paper are available at http://dx.doi.org/10.17630/404e3a8f-3346-4dcc-b90a-62795097b1eb.

References

1. Benedek, N.A.; Fennie, C.J. Hybrid improper ferroelectricity: A mechanism for controllable polarization-magnetization coupling. *Phys. Rev. Lett.* **2011**, *106*, 107204. [CrossRef] [PubMed]

2. Bousquet, E.; Dawber, M.; Stucki, N.; Lichtensteiger, C.; Hermet, P.; Gariglio, S.; Triscone, J.-M.; Ghosez, P. Improper ferroelectricity in perovskite oxide artificial superlattices. *Nature* **2008**, *452*, 732–736. [CrossRef] [PubMed]

3. Benedek, N.A. Origin of Ferroelectricity in a Family of Polar Oxides: The Dion—Jacobson Phases. *Inorg. Chem.* **2014**, *53*, 3769–3777. [CrossRef] [PubMed]

4. Benedek, N.A.; Rondinelli, J.M.; Djani, H.; Ghosez, P.; Lightfoot, P. Understanding ferroelectricity in layered perovskites: New ideas and insights from theory and experiments. *Dalton Trans.* **2015**, *44*, 10543–10558. [CrossRef] [PubMed]

5. Glazer, A.M. The classification of tilted octahedra in perovskites. *Acta Crystallogr.* **1972**, *B28*, 3384–3392. [CrossRef]

6. Woodward, P.M. Octahedral Tilting in Perovskites. I. Geometrical Considerations. *Acta Crystallogr.* **1997**, *B53*, 32–43. [CrossRef]

7. Howard, C.J.; Stokes, H.T. Group-theoretical analysis of octahedral tilting in perovskites. *Acta Crystallogr.* **1998**, *B54*, 782–789. [CrossRef]

8. Oh, Y.S.; Luo, X.; Huang, F.-T.; Wang, Y.; Cheong, S.-W. Experimental demonstration of hybrid improper ferroelectricity and the presence of abundant charged walls in (Ca, Sr)$_3$Ti$_2$O$_7$ crystals. *Nat. Mater.* **2015**, *14*, 407–413. [CrossRef] [PubMed]

9. Li, B.-W.; Osada, M.; Ozawa, T.C.; Sasaki, T. RbBiNb$_2$O$_7$: A New Lead-Free High-T c Ferroelectric. *Chem. Mater.* **2012**, *24*, 3111–3113. [CrossRef]

10. Withers, R.L.; Thompson, J.G.; Rae, A.D. The crystal chemistry underlying ferroelectricity in Bi$_4$Ti$_3$O$_{12}$, Bi$_3$TiNbO$_9$, and Bi$_2$WO$_6$. *J. Solid State Chem.* **1991**, *94*, 404–417. [CrossRef]

11. Snedden, A.; Knight, K.S.; Lightfoot, P. Structural distortions in the layered perovskites CsANb$_2$O$_7$ (A= Nd, Bi). *J. Solid State Chem.* **2003**, *173*, 309–313. [CrossRef]

12. Strayer, M.E.; Gupta, A.S.; Akamatsu, H.; Lei, S.; Benedek, N.A.; Gopalan, V.; Mallouk, T.E. Emergent Noncentrosymmetry and Piezoelectricity Driven by Oxygen Octahedral Rotations in n = 2 Dion–Jacobson Phase Layer Perovskites. *Adv. Funct. Mater.* **2016**, *26*, 1930. [CrossRef]
13. Kumada, N.; Kinomura, N.; Sleight, A.W. $CsLaNb_2O_7$. *Acta Crystallogr.* **1996**, *C52*, 1063–1065. [CrossRef]
14. McCabe, E.E.; Bousquet, E.; Stockdale, C.P.J.; Deacon, C.A.; Tran, T.T.; Halasyamani, P.S.; Stennett, M.C.; Hyatt, N.C. Proper Ferroelectricity in the Dion–Jacobson Material $CsBi_2Ti_2NbO_{10}$: Experiment and Theory. *Chem. Mater.* **2015**, *27*, 8298–8309. [CrossRef]
15. Fennie, C.J.; Rabe, K.M. Ferroelectricity in the Dion-Jacobson $CsBiNb_2O_7$ from first principles. *Appl. Phys. Lett.* **2006**, *88*, 262902. [CrossRef]
16. Goff, R.J.; Keeble, D.; Thomas, P.A.; Ritter, T.; Morrison, F.D.; Lightfoot, P. Leakage and proton conductivity in the predicted ferroelectric $CsBiNb_2O_7$. *Chem. Mater.* **2009**, *21*, 1296–1302. [CrossRef]
17. Chen, C.; Ning, H.; Lepadatu, S.; Cain, M.; Yan, H.; Reece, M.J. Ferroelectricity in Dion–Jacobson $ABiNb_2O_7$ (A= Rb, Cs) compounds. *J. Mater. Chem.* **2015**, *3*, 19–22. [CrossRef]
18. Campbell, B.J.; Stokes, H.T.; Tanner, D.E.; Hatch, D.M. ISODISPLACE: A web-based tool for exploring structural distortions. *J. Appl. Crystallogr.* **2006**, *39*, 607. [CrossRef]
19. Etxebarria, I.; Perez-Mato, J.M.; Boullay, P. The role of trilinear couplings in the phase transitions of Aurivillius compounds. *Ferroelectrics* **2010**, *401*, 17–23. [CrossRef]
20. Snedden, A.; Hervoches, C.H.; Lightfoot, P. Ferroelectric phase transitions in $SrBi_2Nb_2O_9$ and $Bi_5Ti_3FeO_{15}$: A powder neutron diffraction study. *Phys. Rev. B* **2003**, *67*, 092102. [CrossRef]
21. Boullay, P.; Tellier, J.; Mercurio, D.; Manier, M.; Zuñiga, F.J.; Perez-Mato, J.M. Phase transition sequence in ferroelectric Aurivillius compounds investigated by single crystal X-ray diffraction. *Solid State Sci.* **2012**, *14*, 1367–1371. [CrossRef]
22. Larson, A.C.; von Dreele, R.B. *General Structure Analysis System (GSAS)*; Los Alamos National Laboratory Report No. 88-748; Los Alamos National Laboratory: Los Alamos, NM, USA, 1994.
23. Toby, B.H. EXPGUI, a graphical user interface for GSAS. *J. Appl. Crystallogr.* **2001**, *34*, 210–213. [CrossRef]

© 2017 by the authors. Licensee MDPI, Basel, Switzerland. This article is an open access article distributed under the terms and conditions of the Creative Commons Attribution (CC BY) license (http://creativecommons.org/licenses/by/4.0/).

crystals

MDPI

Review

Crystal Structures from Powder Diffraction: Principles, Difficulties and Progress

Radovan Černý

Laboratory of Crystallography, DQMP, University of Geneva, 24 quai Ernest-Ansermet,
CH-1211 Geneva, Switzerland; Radovan.Cerny@unige.ch

Academic Editor: Stevin Pramana
Received: 27 April 2017; Accepted: 11 May 2017; Published: 16 May 2017

Abstract: The structure solution from powder diffraction has undergone an intense evolution during the last 20 years, but is far from being routine. Current challenges of powder crystallography include ab initio crystal structure determination on real samples of new materials with specific microstructures, characterization of intermediate reaction products from in situ, in operando studies and novel phases from in situ studies of phase diagrams. The intense evolution of electron diffraction in recent years, providing an experimental (precession) and theoretical (still under intense development) solution to strong dynamic scattering of electrons, smears the traditional frontier between poly- and single-crystal diffraction. Novel techniques like serial snapshot X-ray crystallography point in the same direction. Finally, for the computational chemistry, theoreticians hand-in-hand with crystallographers develop tools where the theory meets experiment for crystal structure refinement, which becomes an unavoidable step in the validation of crystal structures obtained from powder diffraction.

Keywords: powder diffraction; structure solution; X-ray diffraction; DFT calculations; electron diffraction

1. Crystal Structures from Powders: Where Are We Currently?

Crystal structure determination using powder diffraction (SDPD) started only three years after W.L. Bragg published four crystal structures (NaCl-type) from single crystal data [1]. Peter Debye and Paul Scherrer had solved in 1916 the structure of LiF [2] from X-ray powder diffraction data. The author has no intention to review the long and successful history of SDPD in this review; the reader can consult an excellent review by A.K. Cheetham, Chapter 2 in [3], by W.I.F. David in [4] or by W. Paszkowicz in [5]. The structure solution means in this article determination of the (average) 3D-periodic arrangement of the atoms in the crystal using mainly the information content of Bragg scattering in the diffraction pattern. The case of aperiodic crystals (modulated and quasicrystals), of short-range order (diffuse scattering), as well as of nano-crystals (crystals without measurable Bragg signals) will not be discussed here. The structure prediction, on the other hand, will be used in the sense of the determination of the 3D arrangement of the atoms in the crystal using mainly first-principle calculations, and any experimental observation, like the diffraction pattern, is used only for the validation of the predicted structural model.

1.1. Indexing: Still a Bottleneck

Despite an intense development of indexing algorithms in the last 20 years, the detection of the correct lattice may still be a bottleneck of the structure solution in the case of low resolution powder patterns or in the case of monoclinic and triclinic symmetry. A review of known indexing algorithms may be found in Chapter 7 of [3], Chapter 5 of [6] and Chapter 7 of [7]. Each crystallographer and each crystal may work better with a particular algorithm, but as is said in another excellent indexing review [8]: "Powder indexing works beautifully on good data, but with poor data it will usually not work at all". The author can only agree with that statement and add (unidentified source, but very probably R. Shirley): "Success of the indexing increases proportionally with the number of different applied indexing programs". The author's personal choice is the dichotomy algorithm (see [9] for a historical review) nicely implemented in the programs DICVOL04 [10], Fox [11,12] and X-Cell [13]. The advantage of the dichotomy algorithm is its speed (especially in Fox) and its robustness towards low data quality. Indexing is followed by space group determination, which is based on systematic extinction analysis supported by the knowledge of crystal physical properties, like piezoelectricity, the presence of which excludes centrosymmetric space groups. Fairly robust algorithms for space group determination are based on full pattern fitting (Le Bail or Pawley), available in most software packages.

The indexing can be quite difficult for the crystals lying on opposite ends of the symmetry scale:

Low-symmetry crystals: The indexing algorithm very often proposes for monoclinic crystals several solutions, which are hard to rank on the basis of usual criteria like expected cell volume (based on formula volume and the number of formulas in the cell, in agreement with Wyckoff site multiplicities) and figure-of-merit. A very strong criterion of the cell validity (not only for monoclinic crystals) is the observation of crystallographic extinctions. This means that a cell that shows an extinction of, for example, reflection 020 will rarely be the correct one, unless there is an explanation for it, like the use of X-rays for compounds where hydrogen is not only bonded to a molecule like a terminal ligand, but it is a crystal building element like an anion.

High-symmetry crystals: The indexing algorithms often find a sub-cell, which does not clearly show crystallographic extinctions. The failure of the indexing may not be in an erroneously-used impurity reflection or low precision of peak positions, but in too low of an upper limit for the cell volume. Do not be afraid of big cells! If one of the indexing solutions is a big cell with many crystallographic extinctions pointing clearly to one extinction symbol, it is highly probable that this is the correct one. Very nice examples are recent solutions of two metal borohydrides $KAl(BH_4)_4$ [14] and $Li_3Cs_2(BH_4)_5$ [15], both in the rare space group *Fddd*.

A particular difficulty is indexing of multiphase samples. Even if peaks of known phases are identified in the measured powder pattern, one is never sure that the remaining peaks belong only to one novel phase. An elegant solution, used for a long time in many laboratories, is what the author calls "decomposition-aided indexing" [16]: the in situ powder diffraction data are recorded while the sample is heated up to the disappearing of the peaks of one unknown phase (decomposition, melting, reaction). This allows the separation of the peaks of several novel phases.

As a final word about indexing powder patterns, we may use what the author always says to his students: "If you like your unit cell, then it is correct". This means that all criteria, such as nice crystallographic extinctions, cell volume in agreement with the chemical formula and any relation to the unit cell of related compounds, must be valid simultaneously.

1.2. Fast and Low Noise Data: 2D-Detectors

The 2D-detectors developed for faster and better quality data collection for single crystals became very popular in powder diffraction. The speed of the data collection has opened the door for in situ studies involving fast (few seconds) structural changes (phase transitions, reactions) [17]. In addition to the excellent time resolution, highly accurate diffracted intensities (good powder average) and a lower sensitivity to preferred orientation are obtained by using 2D-detectors as the whole Debye–Scherrer ring is collected and not only a short segment of it determined by the window of the point or linear detectors. This leads to a very low statistical noise in the data (smooth powder pattern), allowing the detection of very weak peaks needed for correct indexing and space group determination. Logically, the next step is curved 2D-detectors, in the ideal case a sphere around the sample allowing high Bragg angles to be collected simultaneously with the low angle data, certainly a challenge for the developers.

1.3. Which Method for Structure Solution?

The structure solution methods may be generally divided into two groups: intensity extraction (IE)-based algorithms working in the reciprocal space (traditionally called reciprocal space methods) and pattern modelling (PM)-based algorithms working in the direct space (traditionally called direct space methods or global optimization methods) and using the chemical knowledge from that space. Hybrid methods iterating between both spaces are known as well, but a decisive criterion to classify a method is whether the method requires integrated intensities of individual peaks in a powder pattern (IE) or not (PM). A schematic view of possible SDPD roads is given in Figure 1. For a comprehensive review of the methods, see [3].

Powder diffraction uses the same methodology for the structure solution regardless of the nature of the compound to be investigated, but the careful selection of the solution algorithm according to the compound may considerably improve the success rate. The separation of compounds by the inorganic/organic boundary is of less importance for the diffraction than the knowledge of how the atoms build up larger building units and the crystal itself. This does not mean a particular difference for IE methods; however, it becomes important for PM. The structure solution algorithm working in the direct space has to know how to define basic structural units (BU) of the crystal, which are then manipulated (optimized) by the algorithm. A molecular/non-molecular boundary is therefore relevant for the choice of a structure determination method.

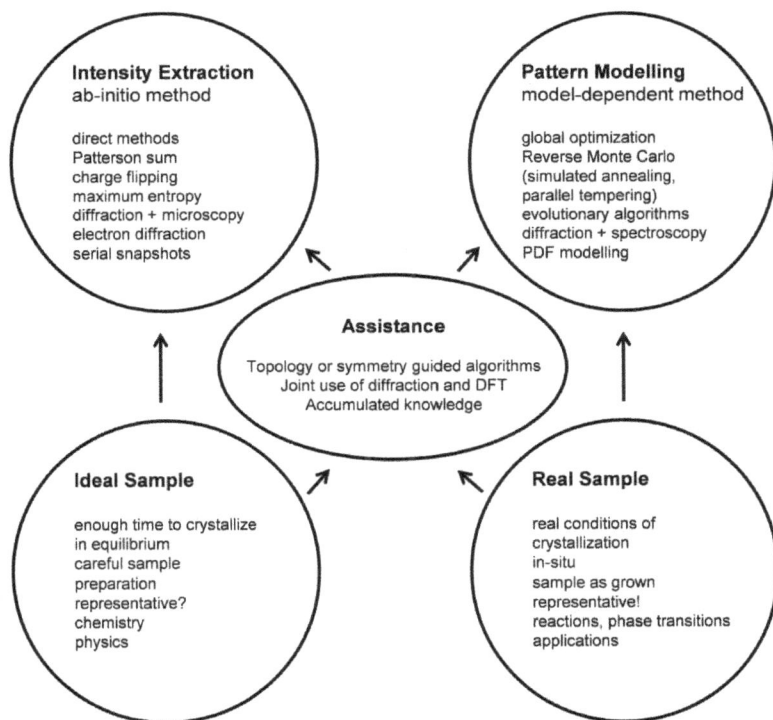

Figure 1. Schema of the structure determination from powder diffraction. While for the ideal sample, there is no need to proceed by assistance of and/or the use of pattern modelling, even if it this road is possible, the real sample has no other choice than to use assisted pattern modelling.

1.3.1. Methods Using Intensity Extraction

The methods that work with the integrated intensities extracted from the powder pattern are the choice when high resolution powder patterns are available, i.e., when an important fraction of reliable integrated intensities may be extracted down to $d \sim 1$ Å. For more specific criteria of what the high resolution pattern means, see Chapter 8 in [3]. Whenever the reliable intensities can be extracted, these methods should be used, because the long time evolution of direct methods (Chapters 10 and 11 in [3]), Patterson sum-based methods (Chapter 13 in [3]), charge-flipping (dual-space method) [18] and maximum entropy method [19], to name the most commonly used, has resulted in powerful tools. The resolution of the powder pattern is nowadays determined mostly by the quality of the sample: the crystals have to have enough time to crystallize, be in thermodynamic equilibrium, and the sample must be carefully prepared [20]. Such a crystal will be certainly used when studying the chemistry or physics of the compound, but the sample is not necessarily representative for a material from applications where, often, out of equilibrium conditions lead to metastable phases.

Several ideas were developed to assist (and improve) the extraction of integrated intensities from the powder pattern: those where no additional data must be collected, like the Patterson function (Chapter 12 in [3]), triplet relation based (Chapter 11 in [3]) or maximum entropy related (Chapter 14 in [3]) and those based on multiple datasets, like texture and thermal dilatation based (Chapter 9 in [3] and [21]).

1.3.2. Methods Using Pattern Modelling

When no reliable integrated intensities can be extracted from a powder pattern, then the only choice is modelling the pattern as a whole, i.e., a synthetic approach contrary to the analytical approach (intensity extraction). Such a situation is typical for real conditions of crystallization as found in many applied materials, in situ studies of reactions and phase transitions. The sample is as grown and is representative for the material. Often, a high time resolution is required to catch the intermediate or rapidly transforming phases. There is no need to extract the integrated intensities; low resolution powder data (~ 2 Å) are often sufficient, and the methods work with patterns containing broadened, overlapped peaks. Any additional information about the atomic coordination and connectivity creating bigger BUs and leading so to a lower number of structural parameters to be determined is easily used. Generally, the pattern modelling methods differ in the algorithm of the global optimization, i.e., the algorithm that globally optimizes the structural model to make its calculated powder pattern better fit the observed one (diffraction cost function). Algorithms are known from the global optimization field of mathematics, and the simplest algorithms, like reversed Monte Carlo (RMC) in simulated annealing and parallel tempering mode or evolution algorithms, are used. For more information, see Chapter 15 in [3] or [22].

PM methods are also easily coupled with any other information aimed to identify the correct crystal structure: knowledge of the chemistry of the unit cell and bonding between the atoms is already actively used in construction and merging of BUs. Crystal energy on the level of electrostatic potentials (ab initio calculations are used for structure validation and prediction, rather than determination) is often coupled with the diffraction cost function (Chapter 15 in [3]). Crystal chemistry contains valuable knowledge accumulated by generations of solid state chemists and crystallographers. While it is mostly based on the observation of known crystal structures and empirical relations among them, it can be very useful in making a decision about which structural model is more probable. The active use of crystallographic databases (see www.iucr.org/resources/data) is highly recommended (author's opinion).

2. Molecular Compounds

A molecule or a complex anion is naturally a BU formed by strong intramolecular interactions (covalent) and packed by intermolecular interactions (dispersive, hydrogen or halogen-bond, π-, coordinating- and ionic-interactions) in the unit cell. No sharing of atoms between the molecules and complex anions occurs. This means that molecular crystals are ready to be easily treated by PM methods. The BU is easily defined when the molecule or complex anion is known. It is probably not a surprise that one of the most complex molecular crystals solved ab initio from powder diffraction data was solved by RMC in parallel tempering mode. It contains 63 organic (non-hydrogen) atoms in the asymmetric unit, and active use of fragments' connectivity knowledge was essential in the structure solution [23].

The most important group of compounds treated with PM methods is probably pharmaceuticals, as they are typically available in powder form, and novel molecules are systematically synthesized. For a review about the powder crystallography of pharmaceuticals, see [24].

3. Non-Molecular Compounds: Extended Solids

Non-molecular compounds, also called extended solids, are constructed by bonds (ionic, covalent) that extend "infinitely" in three dimensions through a crystal. These non-molecular crystals usually crystallize with higher symmetries, and atoms often occupy special Wyckoff positions; building a model for the PM method is therefore not a straightforward procedure [25]. The presence of high-order symmetries leads to a fragmentation of BUs, as any group of atom (e.g., an octahedron) can be located on a symmetry element, so that only part of the atoms of a symmetry-constrained BU are truly independent. The actual arrangement in space of the BUs, relative to the symmetry elements of the unit cell, are a priori unknown. To cope with this, Deem and Newsam used a merging term in the cost function of their PM method [26], which has then evolved in a general and simple algorithm called dynamical occupancy correction (DOC) implemented in the PM program Fox [11].

One of the most important groups of materials that motivated many developments in SDPD is zeolites built up from SiO_4 tetrahedra. For a review, see [27]. Intense development of PM methods was started by zeolite research [26]. The texture method for integrated intensities extraction was developed on zeolite samples [21]. A nice topology-guided dual space method has been developed for zeolites [28], but the approach is easily modified to any other class of compound with a typical underlying topology.

Another group of polyhedral compounds important for SDPD evolution are (among others) oxides. As these are often real materials, PM methods have been intensively applied here. The most complex structure ever solved from powder diffraction (SDPD applied to proteins is usually based on detected isomorphisms) is α-$Bi_2Sn_2O_7$, a 176-atom polyhedral compound, solved by an RMC search among subgroups of the pyrochlore space group *Fd-3m* [29], an approach that can be called the symmetry-guided PM method.

Metal hydrides are rarely available as single crystals due to the method of synthesis and are another SDPD evolution engine [30]. The program Fox has been originally developed on demand from the hydrogen storage community, but has found much broader application [31]. Hydrides are close-packed compounds when formed by hydrogen gas absorption in intermetallic compounds, usually not well crystallized, and therefore, PM methods are used for SDPD. Another group of hydrides, the complex hydrides, contains a homoleptic complex anion with hydrogen as the ligand, a well-studied example being borohydrides. They are prepared typically by mechano-synthesis leading to fine powder excluding IE methods for SDPD. Borohydrides' powder crystallography is a nice example when powder diffraction needs to be assisted by other methods of characterization like various spectroscopies and by crystal chemistry tools (structural analogies between borohydrides and oxides) when facing relatively complex structures [32]. In general, the powder crystallography of metal hydrides is exciting as it provides for all types of crystallographic difficulties, like multi-phase samples, anisotropic line broadening, weak superstructures, pseudo-symmetry, local order in disorder, a weak scatterer close to a strong one, phase transitions and reactions [16].

4. Structure Validation: Help of Theoreticians

Computational chemistry provides information about the studied crystal, which is hardly accessible from powder diffraction. Ab initio calculations on the DFT level can be used at three stages of SDPD:

initial model building (mostly the optimization of a molecule), structure refinement (alternatively with Rietveld refinement) and post-experimental structure validation [33]. The last one helps to validate new structures, locate light hydrogen atoms, especially when using high-pressure diffraction data of limited quality, and even to correct the symmetry and some structural details [15,25,30,34]. Possibly higher crystallographic symmetry of a structure model, which was optimized without any symmetry, can be detected by a suitable algorithm, like the ADDSYM routine in the program Platon [35].

A DFT optimization of the Rietveld-refined structure easily gets away from the local minimum of the experiment cost-function. Since Rietveld refinement and the DFT method are based on two very different cost functions, their combination greatly helps to reach a global minimum. We can therefore highly recommend post-experimental DFT optimization of crystal structures for the systems containing light elements such as hydrogen. One must however keep in mind that if used to validate powder structures, one is employing a far more precise method to "validate" a less precise one. Both may be inaccurate though. Thus, optimizing a structure may not necessarily equate to making it more accurate. Well-founded crystallographic and chemical analysis will always remain indispensable [33].

One inconvenient side-effect of the crystals with a disordered nature like complex hydrides for example is that even experimentally-measured room-temperature structures rarely or never represent the ground state itself. DFT calculations are commonly performed at 0 K where dynamics largely freeze. This obviously severely biases structural predictions by ab initio calculations. For structural corrections/optimizations, the choice needs to be made to either fix the unit cell to experimentally obtained lattice constants (preferably on powder samples) or optimize the cell geometry along with atomic positions. It is up to the user and his/her experience to decide which is more efficient, depending on the specific problem to be addressed.

5. Perspectives

5.1. Electron Crystallography

Electron diffraction has developed rapidly in recent times as a complementary technique with powder diffraction: first, in collaboration with X-ray powder diffraction providing additional information in the phasing process [36] and, nowadays, as a stand-alone technique for ab initio crystal structure solution [37]. The interaction of electrons with the electron density of the crystal being much stronger than that of X-rays means that a single grain (very small single crystal) is enough to produce an exploitable diffraction signal. Having been eliminated from the ab initio structure solution due to dynamic scattering effects for a long time (the dark side of the strong electron-matter interaction), this is nowadays solved by precession electron diffraction, decreasing the dynamic contribution to the kinematic scattering [38]. However, its inherent local sampling on a nanometer scale and substantially complex experimental setup inhibiting in situ coupling to other methods pose serious problems for some experiments. Additionally, the sensitivity of soft matter to the electron beam will limit the method.

5.2. Nuclear Magnetic Resonance Crystallography

Three-dimensional structures of powdered solids can be determined also by combining solid-state NMR spectroscopy, X-ray powder diffraction and DFT calculations. NMR spectroscopy has from its earliest days provided structural information on both periodic and amorphous compounds, ranging from specific internuclear distances to complete structural models of complex materials and biomolecules. The technique "NMR crystallography" is now recognized by the International Union of Crystallography (IUCr) and expected to be an integral element of modern powder crystallography. For more details, see the recent review [39].

5.3. Serial Snapshot X-ray Crystallography

As one of the possible alternatives to powder diffraction, we may notice the serial femtosecond X-crystallography [40]. Similar to electron crystallography, the serial snapshot X-ray crystallography transforms powder diffraction into single crystal diffraction. Contrary to the former, which works with one sub-micrometric single crystal, the latter method accumulates partial single crystal data obtained very quickly on a series of sub-micrometric crystals, thus avoiding any radiation damage. The partial single crystal data are then merged into a complete single crystal dataset. The method is coupled with the pulse nature of novel sources of synchrotron radiation, free electron lasers and is intensively tested on protein crystals [41].

An interesting variant of the serial snapshot crystallography is Laue microdiffraction [42]. A revival of the Laue method for the ab initio structure solution has been observed in the last few years, especially using neutron diffraction. The Laue method allows collecting simultaneously many reflections in one shot and uses the whole spectrum of wavelength available in the primary beam. The serial snapshot version of the Laue method, Laue microdiffraction, works also with a series of sub-microscopic crystals, like serial snapshot crystallography, but thanks to the broad-bandpass mode of some news sources of synchrotron radiation, like the free electron laser constructed at the Paul Scherrer Institute Villigen, Switzerland, with an energy bandwidth of about 4%, the collected data correspond to a series of Laue patterns.

5.4. Pattern Modelling Methods Assisted by Ab Initio Calculations

In spite of great effort invested to the structure prediction from first-principles, the ab initio calculations are far from being a routine tool for predicting the correct crystal structure only from known chemical composition [43]. The prediction of new systems may, however, be a common ground for both crystallography and computational chemistry. Not only can experimentalists provide theoreticians with information for calculations, but predictions on whole systems can also be confirmed by experiment, as was recently illustrated on potassium silanides [44]. Learning from structure predictions is one of the guides that helps the experimentalist to build a model that is then optimized by a global optimizer in a PM method. Such an approach started with close-packed compounds [45]; the greatest progress may be registered among molecular crystals [46]. Learning from crystal structure prediction done, not by ab initio calculations, but by exploiting known topologies and known building units of related crystals, led to the greatest success among framework materials [47–50].

The ideal situation would be naturally the combined use of ab initio calculations with the analysis of diffraction data. Should we speak about the prediction-guided solution or about the solution-guided prediction? It is of no importance who guided whom, only the result, i.e., the correct crystal structure counts. The theoretical and experimental information is easily combined in a global optimizer that searches for the best structural model. Several global optimizers were proposed for structural predictions at the DFT level of the ab initio calculations, i.e., without active use of diffraction information: the evolutionary algorithm [51], simulated annealing [52] and the molecular dynamics-based algorithm (minima hopping) [53]. The greatest problem of the ab initio calculation part is, however, the time needed for the solid state calculations. Without new ideas for how to accelerate the DFT calculations, the global optimization of the structural model using jointly the theory (crystal energy) and experiment (diffraction pattern) will stay limited to structures not going far beyond 100 atoms in the unit cell. One way for improving the situation is distributed computing, i.e., joint use of many computing units: processors in parallel available already nowadays for PM programs like DASH [54] and Fox [12].

5.5. Accumulated Knowledge-Guided Structure Solution

The knowledge accumulated by generations of crystallographers on compounds from a given class of materials must be actively used within the SDPD process. There is no need to reinvent the wheel. Known topologies, crystal chemistry rules, atomic coordination, interatomic distances and angles, molecular fragments, polyhedral connectivity and other information are valuably concentrated in various databases specialized for organic, inorganic compounds, frameworks and many others. It is of high importance that the database allows for a combined search using various parameters:

if during the SDPD process, the reliable indexing provides the unit cell volume and crystal system, chemical analysis provides a reliable chemical composition and the coordination of one or more cations is expected from spectroscopic methods, then the cross search in a database may provide a unique structural prototype, which may immediately lead to the correct structure model for the studied crystal. Such acceleration of the SDPD process cannot be omitted. The utmost care must, however, be paid that the structure solution is the correct one, i.e., not biased and misled to a local minimum in the parameter space of the global optimizer. The wrong conclusions are made easily, and powder diffraction is always limited in its resolving power by the projection of the 3D diffraction pattern on the 1D powder pattern.

6. Conclusions

It is certain that in the future, potentially useful systems in different applications like electroceramics, will involve highly complex and dynamic systems, which are subject to metastability and changing thermodynamic equilibria, particle sizes and crystallinity. The setups at synchrotron beamlines are evolving quickly, and it will become increasingly feasible to study systems under working conditions. The structure solution and complete structural characterization will always be at the forefront of characterizing these systems. Powder diffraction is and will surely remain the most amenable method due to the broad spectrum of information obtainable and the simplicity of in situ experiments coupled to complementary methods, such as X-ray absorption spectroscopy (XAS), vibrational spectroscopy or thermogravimetric analysis (TGA), differential thermal analysis (DTA) and mass spectroscopy (MS), just to name the most useful ones.

Acknowledgments: Author acknowledge the hard work of all colleagues solving the crystal structures from powder diffraction data. Special thank belongs to those using the program Fox.

Conflicts of Interest: The author declares no conflict of interest. The founding sponsors had no role in the design of the study; in the collection, analyses, or interpretation of data; in the writing of the manuscript, and in the decision to publish the results.

References

1. Bragg, W.L. The structure of some crystals as indicated by their diffraction of X-rays. *Proc. R. Soc. Lond.* **1913**, *A89*, 248–277. [CrossRef]
2. Debye, P.; Scherrer, P. Interferenzen an regellos orientierten Teilchen im Röntgenlicht. *Phys. Z.* **1916**, *17*, 277.
3. David, W.I.F.; Shankland, K.; McCusker, L.B.; Baerlocher, C. *Structure Determination from Powder Diffraction Data*; Oxford University Press: Oxford, UK, 2002.
4. David, W.I.F. Power of Powder Diffraction, Int. School of Crystallography, Erice, pages 19–29. Available online: www.iucr.org/resources/commissions/powder-diffraction/schools/erice2011 (accessed on 2 June 2011).
5. Paszkowicz, W. Ninety years of powder diffraction: From birth to maturity. *Synchrotron Radiat. Nat. Sci.* **2006**, *5*, 1–2.
6. Pecharsky, V.K.; Zavalij, P.Y. *Fundametals Of Powder Diffraction and Structural Characterization of Materials*; Kluwer Academic Publishers: Norwell, MA, USA, 2003.
7. Dinnebier, R.E.; Billinge, S.J.L. *Powder Diffraction: Theory and Practice*; The Royal Society of Chemistry: Cambridge, UK, 2008.
8. Bergmann, J.; Le Bail, A.; Shirley, R.; Zlokazov, V. Renewed interest in powder diffraction data indexing. *Z. Kristalogr.* **2004**, *219*, 783–790. [CrossRef]
9. Boultif, A. History of dichotomy method for powder pattern indexing. *Powder Diffr.* **2005**, *20*, 284–287. [CrossRef]
10. Boultif, A.; Louër, D. Powder pattern indexing with the dichotomy method. *J. Appl. Cryst.* **2004**, *37*, 724–731. [CrossRef]
11. Favre-Nicolin, V.; Černý, R. "Free objects for crystallography": A modular approach to ab initio structure determination from powder diffraction. *J. Appl. Cryst.* **2002**, *35*, 734–743. [CrossRef]
12. Černý, R.; Favre-Nicolin, V.; Rohlícek, J.; Hušák, M.; Matej, Z.; Kužel, R. Expanding FOX: Auto-indexing, grid computing, profile fitting. *CPD Newsl.* **2007**, *35*, 16–19.

13. Neumann, M.A. X-Cell: A novel indexing algorithm for routine tasks and difficult cases. *J. Appl. Cryst.* **2003**, *36*, 356–365. [CrossRef]
14. Dovgaliuk, I.; Ban, V.; Sadikin, Y.; Černý, R.; Aranda, L.; Casati, N.; Devillers, M.; Filinchuk, Y. The first halide-free bimetallic aluminium borohydride: Synthesis, structure, stability and the decomposition pathway. *J. Phys. Chem.* **2014**, *118*, 145–153. [CrossRef]
15. Schouwink, P.; Smrčok, L.; Černý, R. The role of the Li⁻ node in the Li-BH₄ substructure of double-cation tetrahydroborates. *Acta Cryst. B* **2014**, *70*, 871–878.
16. Černý, R.; Filinchuk, Y. Complex inorganic structures from powder diffraction: Case of tetrahydroborates of light metals. *Z. Krist.* **2011**, *226*, 882–891. [CrossRef]
17. Fitch, A.; Curfs, C. The Power of Powder Diffraction, Int. School of Crystallography, Erice, pages 103–112. Available online: www.iucr.org/resources/commissions/powder-diffraction/schools/erice2011 (accessed on 2 June 2011).
18. Palatinus, L. The Power of Powder Diffraction, Int. School of Crystallography, Erice, pages 160–169. Available online: www.iucr.org/resources/commissions/powder-diffraction/schools/erice2011 (accessed on 2 June 2011).
19. Gilmore, C.J. The Power of Powder Diffraction, Int. School of Crystallography, Erice, pages 142–159. Available online: www.iucr.org/resources/commissions/powder-diffraction/schools/erice2011 (accessed on 2 June 2011).
20. Shankland, K. The Power of Powder Diffraction, Int. School of Crystallography, Erice, pages 37–42. Available online: www.iucr.org/resources/commissions/powder-diffraction/schools/erice2011 (accessed on 2 June 2011).
21. Baerlocher, C.; McCusker, L.B.; Prokic, S.; Wessels, T. Exploiting texture to estimate the relative intensities of overlapping reflections. *Z. Krist.* **2004**, *219*, 803–812. [CrossRef]
22. Černý, R.; Favre-Nicolin, V. Direct space methods of structure determination from powder diffraction: Principles, guidelines and perspectives. *Z. Krist.* **2007**, *222*, 105–113. [CrossRef]
23. Kwon, S.; Shin, H.S.; Gong, J.; Eom, J.H.; Jeon, A.; Yoo, S.H.; Chung, I.S.; Cho, S.J.; Lee, H.S. Self-assembled peptide architecture with a tooth shape: Folding into shape. *J. Am. Chem. Soc.* **2011**, *133*, 17618–17621. [CrossRef] [PubMed]
24. Harris, K.M.D. Powder diffraction crystallography of molecular solids. *Top. Curr. Chem.* **2012**, *315*, 133–178. [PubMed]
25. Černý, R. The Power of Powder Diffraction, Int. School of Crystallography, Erice, pages 62–69. Available online: www.iucr.org/resources/commissions/powder-diffraction/schools/erice2011 (accessed on 2 June 2011).
26. Deem, M.W.; Newsam, J.M. Determination of 4-connected framework crystal structures by simulated annealing. *Nature* **1989**, *342*, 260–262. [CrossRef]
27. Burton, A.W. Structure solution of zeolites from powder diffraction data. *Z. Krist.* **2004**, *219*, 866–880. [CrossRef]
28. Grosse-Kunstleve, R.W.; McCusker, L.B.; Baerlocher, C. Zeolite structure determination from powder diffraction data: Applications of the FOCUS method. *J. Appl. Cryst.* **1999**, *32*, 536–542. [CrossRef]
29. Evans, I.R.; Howard, J.A.K.; Evans, J.S.O. α-Bi₂Sn₂O₇—A 176 atom crystal structure from powder diffraction data. *J. Mater. Chem.* **2003**, *13*, 2098–2103. [CrossRef]
30. Černý, R. Solving crystal structures of metal and chemical hydrides. *Z. Krist.* **2008**, *223*, 607–616. [CrossRef]
31. Favre-Nicolin, V.; Černý, R. A better FOX: Using flexible modelling and maximum likelihood to improve direct-space ab initio structure determination from powder diffraction. *Z. Krist.* **2004**, *219*, 847–856. [CrossRef]
32. Schouwink, P.; Černý, R. Complex hydrides—When powder diffraction needs help. *Chimia* **2014**, *1/2*, 38–44. [CrossRef] [PubMed]
33. Smrčok, L. The Power of Powder Diffraction, Int. School of Crystallography, Erice, pages 231–238. Available online: www.iucr.org/resources/commissions/powder-diffraction/schools/erice2011 (accessed on 2 June 2011).
34. Van de Streek, J.; Neumann, M.A. Validation of molecular crystal structures from powder diffraction data with dispersion-corrected density functional theory (DFT-D). *Acta Cryst.* **2014**, *B70*, 1020–1032. [CrossRef] [PubMed]
35. Spek, A.L. Structure validation in chemical crystallography. *Acta Cryst.* **2009**, *D65*, 148–155. [CrossRef] [PubMed]

36. McCusker, L.B.; Baerlocher, C. Electron crystallography as a complement to X-ray powder diffraction techniques. *Z. Krist.* **2013**, *228*, 1–10. [CrossRef]

37. Kolb, U. The Power of Powder Diffraction, Int. School of Crystallography, Erice, pages 30–36. Available online: www.iucr.org/resources/commissions/powder-diffraction/schools/erice2011 (accessed on 2 June 2011).

38. Zou, X.D.; Hovmöller, S.; Oleynikov, P. *Electron Crystallography—Electron Microscopy and Electron Diffraction*; Oxford University Press: Oxford, UK, 2011.

39. Bryce, D.L.; Taulelle, F. NMR crystallography. *Acta Cryst.* **2017**, *C73*, 126–127. [CrossRef] [PubMed]

40. Chapman, H.N.; Fromme, P.; Barty, A.; White, T.A.; Kirian, R.A.; Aquila, A.; Hunter, M.S.; Schulz, J.; DePonte, D.P.; Weierstall, U.; et al. Femtosecond X-ray protein nanocrystallography. *Natrue* **2011**, *470*, 73–77. [CrossRef] [PubMed]

41. Boutet, S.; Lomb, L.; Williams, G.J.; Barends, T.R.; Aquila, A.; Doak, R.B.; Weierstall, U.; DePonte, D.P.; Steinbrener, J.; Shoeman, R.L.; et al. High-resolution protein structure determination by serial femtosecond crystallography. *Science* **2012**, *337*, 362–364. [CrossRef] [PubMed]

42. Dejoie, C.; McCusker, L.B.; Baerlocher, C.; Abela, R.; Patterson, B.D.; Kunz, M.; Tamura, N. Using a non-monochromatic microbeam for serial snapshot crystallography. *J. Appl. Cryst.* **2013**, *46*, 791–794. [CrossRef]

43. Woodley, S.M.; Catlow, R. Crystal structure prediction from first principles. *Nat. Mater.* **2008**, *7*, 937–946. [CrossRef] [PubMed]

44. Chotard, J.N.; Tang, W.S.; Raybaud, P.; Janot, R. Potassium silanide ($KSiH_3$): A reversible hydrogen storage material. *Chem. Eur. J.* **2011**, *17*, 12302–12309. [CrossRef] [PubMed]

45. Pannetier, J.; Bassas-Alsina, J.; Rodriguez-Carvajal, J.; Caignaert, V. Prediction of crystal structures from crystal chemistry rules by simulated annealing. *Natrue* **1990**, *346*, 343–345. [CrossRef]

46. Neumann, M.A.; Leusen, F.J.J.; Kendrick, J. A major advance in crystal structure prediction. *Angew. Chem. Int. Ed.* **2008**, *47*, 2427–2430. [CrossRef] [PubMed]

47. Mellot Draznieks, C.; Newsam, J.M.; Gorman, A.M.; Freeman, C.M.; Férey, G. De novo prediction of inorganic structures developed through automated assembly of secondary building units (AASBU Method). *Angew. Chem. Int. Ed.* **2000**, *39*, 2270–2275. [CrossRef]

48. Foster, M.D.; Simperler, A.; Bell, R.G.; Delgado Friedrichs, O.; Almeida Paz, F.A.; Klinowski, J. Chemically feasible hypothetical crystalline networks. *Nat. Mater.* **2004**, *3*, 234–238. [CrossRef] [PubMed]

49. Le Bail, A. Inorganic structure prediction with GRINSP. *J. Appl. Cryst.* **2005**, *38*, 389–395. [CrossRef]

50. Morris, W.; Volosskiy, B.; Demir, S.; Gándara, F.; McGrier, P.L.; Furukawa, H.; Cascio, D.; Stoddart, J.F.; Yaghi, O.M. Synthesis, structure, and metalation of two new highly porous zirconium metal–organic frameworks. *Inorg. Chem.* **2012**, *51*, 6443–6445. [CrossRef] [PubMed]

51. Oganov, A.R.; Glass, C.W. Crystal structure prediction using ab initio evolutionary techniques: Principles and applications. *J. Chem. Phys.* **2006**, *124*, 244704. [CrossRef] [PubMed]

52. Schön, J.C.; Pentin, I.V.; Jansen, M. Ab initio computation of low-temperature phase diagrams exhibiting miscibility gaps. *Phys. Chem. Chem. Phys.* **2006**, *8*, 1778–1784. [CrossRef] [PubMed]

53. Goedecker, S. Minima hopping: An efficient search method for the global minimum of the potential energy surface of complex molecular systems. *J. Chem. Phys.* **2004**, *120*, 9911–9917. [CrossRef] [PubMed]

54. David, W.I.F.; Shankland, K.; Van de Streek, J.; Pidcock, E.; Motherwell, W.D.S.; Cole, J.C. DASH: A program for crystal structure determination from powder diffraction data. *J. Appl. Cryst.* **2006**, *39*, 910–915. [CrossRef]

© 2017 by the author. Licensee MDPI, Basel, Switzerland. This article is an open access article distributed under the terms and conditions of the Creative Commons Attribution (CC BY) license (http://creativecommons.org/licenses/by/4.0/).

crystals

MDPI

Article

Investigation into the Effect of Sulfate and Borate Incorporation on the Structure and Properties of SrFeO$_{3-\delta}$

Abbey Jarvis and Peter Raymond Slater *

School of Chemistry, University of Birmingham, Birmingham B15 2TT, UK; axj278@student.bham.ac.uk
* Correspondence: p.r.slater@bham.ac.uk; Tel.: +44-121-414-8906

Academic Editor: Stevin Pramana
Received: 28 April 2017; Accepted: 3 June 2017; Published: 7 June 2017

Abstract: In this paper, we demonstrate the successful incorporation of sulfate and borate into SrFeO$_{3-\delta}$, and characterise the effect on the structure and conductivity, with a view to possible utilisation as a cathode material in Solid Oxide Fuel Cells. The incorporation of low levels of sulfate/borate is sufficient to cause a change from a tetragonal to a cubic cell. Moreover, whereas heat treatment of undoped SrFeO$_{3-\delta}$ under N$_2$ leads to a transformation to brownmillerite Sr$_2$Fe$_2$O$_5$ with oxygen vacancy ordering, the sulfate/borate-doped samples remain cubic under the same conditions. Thus, sulfate/borate doping appears to be successful in introducing oxide ion vacancy disorder in this system.

Keywords: solid oxide fuel cell; cathode; perovskite; sulfate; borate

1. Introduction

Research into solid oxide fuel cells (SOFCs) as alternate energy materials has grown due to their high efficiency and consequent reduction in greenhouse gas emissions. Specifically, research into perovskite materials has attracted significant interest for potential SOFC materials including cathode, electrolyte and anode materials (see, for example, the review articles [1,2]). Traditionally, doping strategies for perovskite materials has involved the introduction of cations of a similar size e.g., Sr^{2+} for La^{3+}, Mg^{2+} for Ga^{3+} to give the oxide ion conducting electrolyte La$_{1-x}$Sr$_x$Ga$_{1-y}$Mg$_y$O$_{3-x/2-y/2}$ [3]. More recently, we have applied oxyanion (silicate, phosphate, sulfate, borate) doping strategies to improve the properties of solid oxide fuel cell materials, initially demonstrating the successful incorporation into Ba$_2$In$_2$O$_5$ [4]. Research on oxyanion incorporation into perovskites was initially reported for superconducting cuprate materials. This work showed that a wide range of oxyanions (carbonate, borate, nitrate, sulfate, and phosphate) could be incorporated into the perovskite structure [5–12]. In this doping strategy, the central ion of the oxyanion group is located on the perovskite B cation site, with the oxygens of this group occupying either three (borate, carbonate, nitrate) or four (sulfate, phosphate) of the available six anion positions around this site, albeit suitably displaced to achieve the correct coordination for the oxyanion group. As an extension to this work on cuprate superconductors, we have demonstrated successful oxyanion doping of perovskite materials for SOFC applications. This has included oxyanion (sulfate, silicate, phosphate) doping of Ba$_2$(In/Sc)$_2$O$_5$ electrolyte materials [13–16]. Doping with these oxyanions introduces disorder on the oxygen sublattice, leading to improvements in the ionic conductivity. This work has also been extended to potential solid oxide fuel cell electrode materials, with successful oxyanion doping in the perovskite systems SrCoO$_3$ [17–20], SrCo$_{0.85}$Fe$_{0.15}$O$_3$ [21], SrMnO$_3$ [18], CaMnO$_3$ [22] and SrFeO$_{3-\delta}$ [23], with the results suggesting improved performance/stability.

Following the prior work demonstrating the successful incorporation of silicate into $SrFeO_{3-\delta}$ [23], we report in this paper an investigation into the effect of sulfate and borate doping. The undoped system $SrFeO_{3-\delta}$ has attracted interest due to its high ionic and electronic conductivity, however under low $p(O_2)$ the oxide ion (and electronic conductivity) are significantly reduced. This is due to oxygen loss leading to a transformation to the oxygen vacancy ordered brownmillerite structure, $Sr_2Fe_2O_5$ (Figure 1). We have therefore investigated the effect of borate/sulfate doping on this transformation.

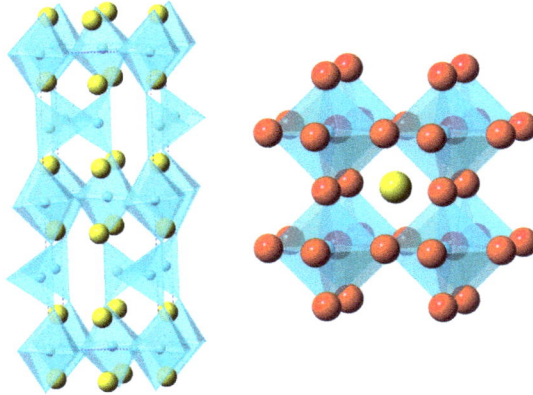

Figure 1. Structure of $Sr_2Fe_2O_5$ (showing oxygen vacancy ordering) (**left**) and $SrFeO_3$ (**right**).

2. Results and Discussion

2.1. $SrFe_{1-x}S_xO_{3-\delta}$

2.1.1. X-ray Diffraction Data

A range of samples of $SrFe_{1-x}S_xO_{3-\delta}$ with increasing sulfate content ($0 \leq x \leq 0.1$) were prepared. X-ray diffraction analysis showed that without sulfate doping $SrFeO_{3-\delta}$ forms a tetragonal perovskite in line with prior reports. This is illustrated in Figure 2 (expanded region $2\theta = 45°$ to $60°$) where peak splitting can clearly be observed. Upon doping with sulfate, there is a transformation to a cubic cell ($0.025 \leq x \leq 0.075$), where no peak splitting is now observed (Figure 2). Above $x = 0.075$, small $SrSO_4$ impurities appear, suggesting that the solubility limit of sulfate in $SrFe_{1-x}S_xO_{3-\delta}$ is $x \approx 0.075$.

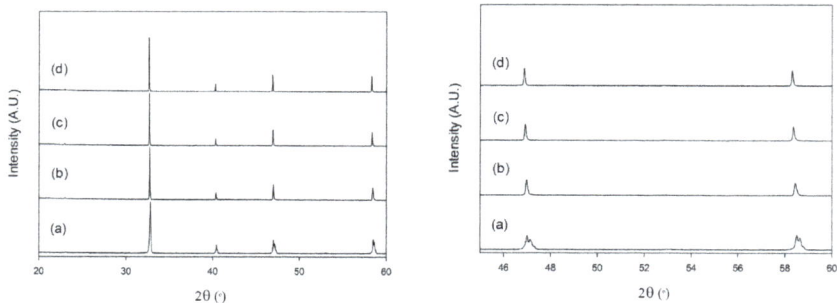

Figure 2. *Cont.*

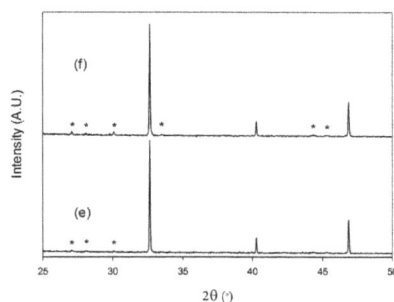

Figure 2. X-ray diffraction patterns for (**a**) $SrFeO_{3-\delta}$, (**b**) $SrFe_{0.975}S_{0.025}O_{3-\delta}$, (**c**) $SrFe_{0.95}S_{0.05}O_{3-\delta}$, (**d**) $SrFe_{0.925}S_{0.075}O_{3-\delta}$, (**e**) $SrFe_{0.9}S_{0.1}O_{3-\delta}$, and (**f**) $SrFe_{0.85}S_{0.15}O_{3-\delta}$. The data show a tetragonal cell for $SrFeO_{3-\delta}$, (as highlighted by the peak splitting: see expanded region Figure $2\theta = 45°$ to $60°$), while the sulfate-doped samples are cubic. $SrSO_4$ impurities (highlighted by an asterisk) are observed for samples with higher sulfate contents (x = 0.1 and 0.15).

In order to provide further support for the incorporation of sulfate, equivalent Fe-deficient samples were prepared without addition of sulfate. The X-ray diffraction data for $SrFe_{0.95}O_{3-\delta}$ (Fe-deficient, no sulfate) and $SrFe_{0.95}S_{0.05}O_{3-\delta}$ are compared in Figure 3. These data show the presence of impurities for $SrFe_{0.95}O_{3-\delta}$ along with peak splitting, consistent with a tetragonal cell, as for the undoped $SrFeO_{3-\delta}$ parent phase. In contrast, impurities are not observed for the sulfate-containing phase $SrFe_{0.95}S_{0.05}O_{3-\delta}$, and the cell is now cubic. Therefore, this comparison provides further evidence for the successful incorporation of sulfate.

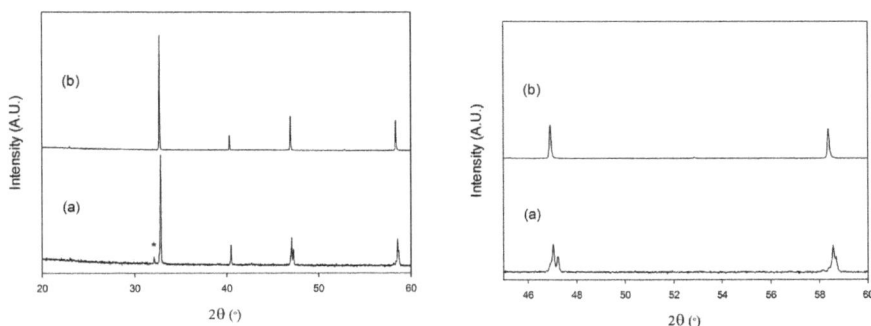

Figure 3. X-ray diffraction patterns for (**a**) $SrFe_{0.95}O_{3-\delta}$ (Fe-deficient, no sulfate) and (**b**) $SrFe_{0.95}S_{0.05}O_{3-\delta}$. The data show impurities (highlighted by an asterisk) and a tetragonal cell for $SrFe_{0.95}O_{3-\delta}$, while the sulfate-containing sample $SrFe_{0.95}S_{0.05}O_{3-\delta}$ is phase pure and cubic.

Cell parameters for all these phases were determined using the Rietveld method (an example fit is shown in Figure 4). The variation of the cell parameters for $SrFe_{1-x}S_xO_{3-\delta}$ with increasing sulfate content is given in Table 1. The data show a small general increase with increasing sulfate content, and this will be discussed in more detail in Section 2.1.4.

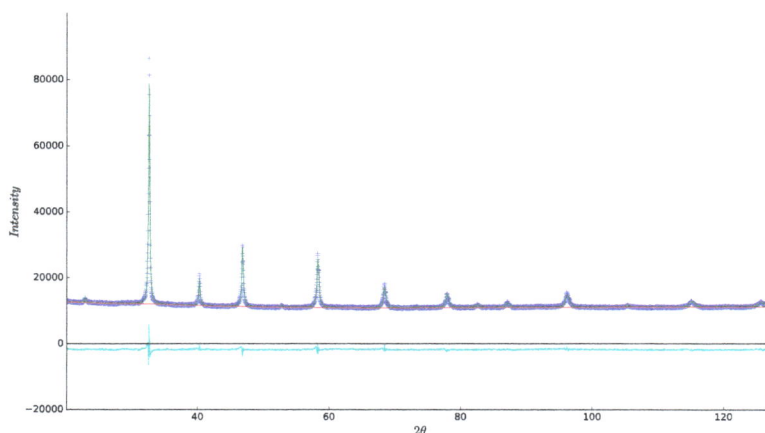

Figure 4. Observed, calculated and difference X-ray diffraction profile for $SrFe_{0.975}S_{0.025}O_{3-\delta}$.

In addition to the determination of the unit cell parameters, site occupancies were refined for the Fe/S site. These occupancies are given in Table 1, and show that the refined values are in good agreement with the expected values.

Table 1. Lattice parameters and Fe/S site occupancies obtained from Rietveld refinement using XRD data for $SrFe_{1-x}S_xO_{3-\delta}$. The structure of $SrFeO_{3-\delta}$ was refined using a tetragonal space group (P4/mmm). The structures of the sulfate-doped samples were refined using a cubic space group (Pm$\bar{3}$m).

	$SrFe_{1-x}S_xO_{3-\delta}$			
S (x)	0	0.025	0.05	0.075
a (Å)	3.8648(1)	3.8723(1)	3.8776(1)	3.8766(1)
c (Å)	3.8487(1)	-	-	-
V (Å^3)	57.486(4)	58.066(2)	58.303(4)	58.260(4)
R_{wp} (%)	1.84	1.67	2.01	1.97
R_{exp} (%)	0.92	0.92	0.90	0.90
Fe occupancy	1	0.98(1)	0.96(1)	0.94(1)
S occupancy	-	0.02(1)	0.04(1)	0.06(1)
Fe/S Uiso	0.003(1)	0.009(1)	0.011(1)	0.008(1)

2.1.2. Stability under N_2

The effect of heating the $SrFe_{1-x}S_xO_{3-\delta}$ samples under N_2 was then examined. Upon heating under N_2 to 950 °C, the X-ray diffraction data showed that $SrFeO_{3-\delta}$ transforms to the oxygen vacancy ordered brownmillerite type $Sr_2Fe_2O_5$ (Figures 1 and 5). This would be expected to be unfavorable for fuel cell applications due to the ordering of oxygen vacancies, which is expected to lower the oxide ion conductivity. In contrast, for the sulfate-doped samples, the disordered cubic perovskite is retained under a similar heat treatment. For the x = 0.025 sample, there are some very weak peaks (see expanded XRD figure) associated with the brownmillerite structure, but for the higher sulfate contents (x ≥ 0.05), a single phase cubic cell is observed. In line with the reduction in the Fe oxidation state towards 3+, there is an increase in the cell parameters associated with the larger size of Fe^{3+} versus Fe^{4+} (Table 2).

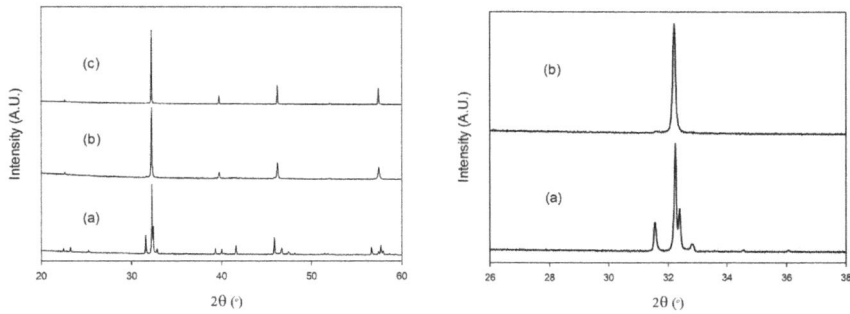

Figure 5. X-ray diffraction patterns of (**a**) SrFeO$_{3-\delta}$, (**b**) SrFe$_{0.975}$S$_{0.025}$O$_{3-\delta}$, and (**c**) SrFe$_{0.95}$S$_{0.05}$O$_{3-\delta}$ after heating under N$_2$ to 950 °C.

Table 2. Lattice parameters for SrFe$_{1-x}$S$_x$O$_{3-\delta}$ after heating in air and N$_2$.

	\multicolumn{6}{c}{**SrFe$_{1-x}$S$_x$O$_{3-\delta}$**}					
S (x)	\multicolumn{2}{c}{0.025}		\multicolumn{2}{c}{0.05}		\multicolumn{2}{c}{0.075}
	Air	Dry N$_2$	Air	Dry N$_2$	Air	Dry N$_2$
a (Å)	3.8723(1)	3.9231(1)	3.8776(1)	3.9256(1)	3.8766(1)	3.9280(1)
c (Å)	-	-	-	-	-	-
V (Å^3)	58.066(2)	60.379(1)	58.303(4)	60.496(1)	58.260(4)	60.606(1)
R$_{wp}$ (%)	1.67	3.10	2.01	3.09	1.97	3.20
R$_{exp}$ (%)	0.92	2.59	0.90	2.51	0.90	2.50

2.1.3. Thermogravimetric Analysis

Thermogravimetric analysis (TGA) (heat treatment in N$_2$ to reduce the Fe oxidation state to 3+) was then utilized to determine the oxygen contents of the samples as prepared in air. This analysis resulted in an interesting observation associated with Mass spectrometry analysis of the evolved gas. For the undoped sample SrFeO$_{3-\delta}$, the loss of mass is only associated with oxygen as indicated by the mass spectrometry data. However, for the sulfate-doped samples, a mass loss associated with CO$_2$ was observed in addition to the expected mass loss due to O$_2$.

Thus, the results indicated the presence of some carbonate in the sulfate-doped samples. In order to remove this carbonate, heat treatment in O$_2$ (up to 900 °C) was carried out for these SrFe$_{1-x}$S$_x$O$_{3-\delta}$ samples. After this oxygen treatment, TGA analysis indicated no presence of carbonate. This is illustrated in Figure 6, where a mass loss associated with CO$_2$ is not observed after heat treatment in O$_2$ for SrFe$_{0.95}$S$_{0.05}$O$_{3-\delta}$.

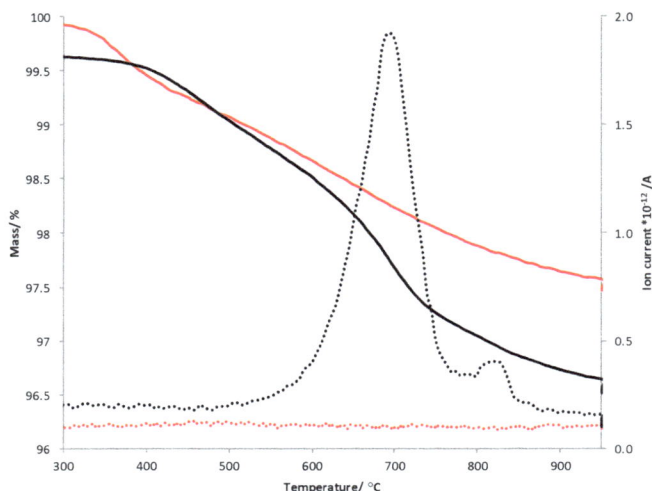

Figure 6. Plot of mass vs. temperature and ion current (for $m/z = 44$; CO_2) vs. temperature (under N_2) for $SrFe_{0.95}S_{0.05}O_{3-\delta}$ prepared in air (black) and O_2 (red), showing a mass loss associated with CO_2 in the air synthesised sample, which is eliminated after heat treatment in O_2. Solid lines indicate %mass and dashed lines indicate ion current.

The TGA results indicating the presence of carbonate may at first glance suggest the existence of a small amount of $SrCO_3$ impurity. However, the temperature at which the CO_2 is lost is significantly lower than would be expected for $SrCO_3$. In order to illustrate this, TGA data for $SrCO_3$ were also collected and compared to the data for the $SrFe_{0.95}S_{0.05}O_{3-\delta}$. This experiment shows a significant difference in the temperature at which the loss of CO_2 occurs for $SrFe_{0.95}S_{0.05}O_{3-\delta}$ and $SrCO_3$. In particular, the starting temperature for CO_2 loss for $SrFe_{0.95}S_{0.05}O_{3-\delta}$ occurs at a significantly lower temperature ($\approx$490 °C vs. $\approx$760 °C for $SrCO_3$), which may suggest that this carbonate is present in the perovskite structure, i.e., we have a mixed sulfate/carbonate-doped sample—$SrFe_{1-x-y}S_xC_yO_{3-\delta}$. Further work is required to investigate this possibility, although as noted in the introduction, carbonate has been shown previously to be accommodated in perovskite materials.

Given the dual (O_2 and CO_2) mass loss for the air-synthesized $SrFe_{1-x}S_xO_{3-\delta}$ samples, it is not possible to determine reliable oxygen contents for these samples. In addition, as detailed by Starkov et al., the determination of oxygen contents in partially substituted ferrites is non-trivial without a reliable fixed reference point [24]. Since it is possible that there may still be a small amount of Fe^{4+} in these samples after the N_2 treatment, there is not a conclusive fixed reference oxygen point, and so any calculated oxygen contents would only be rough approximations.

2.1.4. Heat Treatment under O_2

The samples (described above) heated under O_2 were also examined by X-ray diffraction (Figure 7). All the sulfate-doped $SrFe_{1-x}S_xO_{3-\delta}$ samples were shown to retain their original cubic structure while undoped $SrFeO_{3-\delta}$ remained tetragonal.

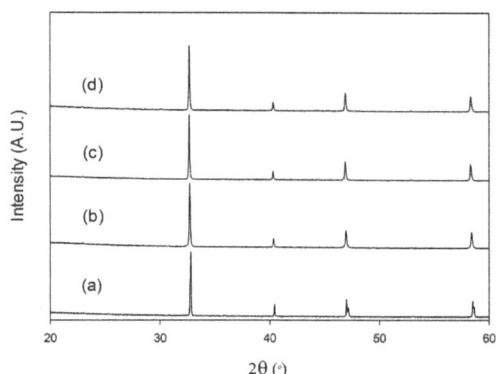

Figure 7. X-ray diffraction patterns of (**a**) SrFeO$_{3-\delta}$, (**b**) SrFe$_{0.975}$S$_{0.025}$O$_{3-\delta}$, (**c**) SrFe$_{0.95}$S$_{0.05}$O$_{3-\delta}$, and (**d**) SrFe$_{0.925}$S$_{0.075}$O$_{3-\delta}$ after heating under O$_2$.

Cell parameters and site occupancies were determined by Rietveld refinement using the X-ray diffraction data. From these structure refinements (Table 3), there is a decrease in the unit cell parameters upon heating in O$_2$. This can be more clearly seen in Figure 8 where the variation in cell volume versus sulfate content is compared for samples heated in O$_2$ and those just heated in air. These data show a reduction in cell volume on heating in oxygen, which can be correlated with a greater concentration of the smaller Fe^{4+} as a result of an increase in oxygen content. The cell parameter data also show an interesting variation with sulfate content, with an approximately linear increase in cell volume up to x = 0.05. The fact that there is no further increase for x = 0.075 may therefore suggest that the sulfate solubility limit is closer to x = 0.05.

The increase in cell volume with increasing sulfate content is at first glance unexpected given that S^{6+} is significantly smaller than Fe$^{3+/4+}$. One possible explanation could relate to changes in the Fe^{3+}/Fe^{4+} ratio on sulfate incorporation. However, this cell volume increase is also seen for the N$_2$ treated samples (Table 2), where we only have Fe^{3+}, and thus another factor must be significant. In this respect, the cell volume increase may be associated with the extra oxygen associated with this dopant. Thus, if we take the case of the N$_2$ treated samples, we are effectively replacing Fe(III)O$_{1.5}$ with SO$_3$, which means the introduction of an extra 1.5 oxide ions per sulfate, which might be expected to contribute to an expanded cell size.

Table 3. Lattice parameters and Fe/S site occupancies for SrFe$_{1-x}$S$_x$O$_{3-\delta}$, obtained from Rietveld refinement using XRD data for SrFe$_{1-x}$S$_x$O$_{3-\delta}$ after heating in O$_2$.

SrFe$_{1-x}$S$_x$O$_{3-\delta}$ Heated in O$_2$			
S (x)	0	0.025	0.05	0.075
a (Å)	3.8651(1)	3.8641(1)	3.8692(1)	3.8691(1)
c (Å)	3.8477(1)	-	-	-
V (Å^3)	57.349(3)	57.694(2)	57.924(2)	57.922(3)
R$_{wp}$ (%)	4.16	3.29	3.79	3.93
R$_{exp}$ (%)	3.71	2.81	2.73	2.79
Fe occ	1(-)	0.97(2)	0.94(2)	0.93(2)
S occ	-	0.03(2)	0.06(2)	0.07(2)

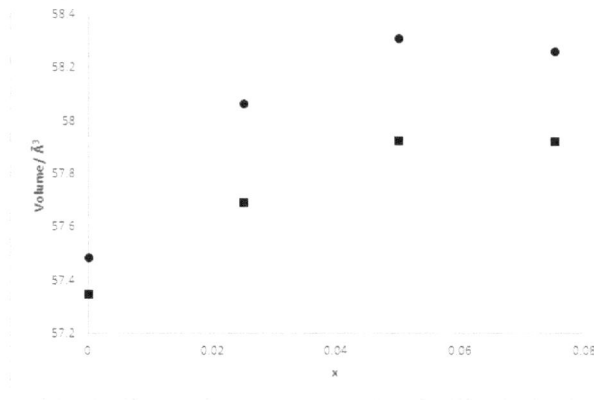

Figure 8. Plot of unit cell volume vs. x for $SrFe_{1-x}S_xO_{3-\delta}$ heated in air (●) and heated in O_2 (■).

2.1.5. Conductivity Data

Following the successful incorporation of sulfate, the conductivities of the $SrFe_{1-x}S_xO_{3-\delta}$ samples were examined in air. In general, the $SrFe_{1-x}S_xO_{3-\delta}$ samples were found to have similar conductivities, with the exception of a notable decrease at lower temperatures for the higher sulfate content (x = 0.075) sample (Figure 9). This observation of a decrease in conductivity for higher dopant levels is comparable to the silicon-doped $SrFeO_{3-\delta}$ system where it was proposed that at higher doping levels the silicate disrupts the Fe-O network resulting in a decrease in conductivity. [23] Another factor, however, could relate to low levels of insulating impurities in this high sulfate content sample, given that the cell parameter data suggest that the sulfate solubility limit may be closer to x = 0.05.

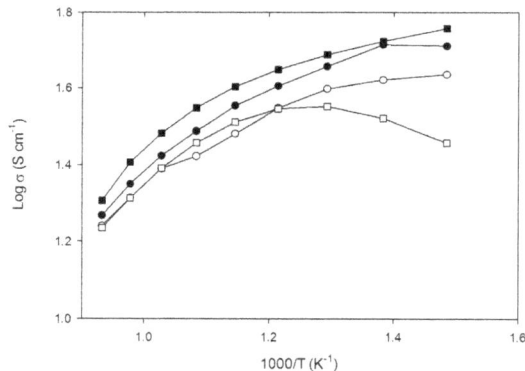

Figure 9. Plot of log σ vs. 1000/T for $SrFeO_{3-\delta}$ (●), $SrFe_{0.975}S_{0.025}O_{3-\delta}$ (○), $SrFe_{0.95}S_{0.05}O_{3-\delta}$ (■) and $SrFe_{0.925}S_{0.075}O_{3-\delta}$ (□) in air.

2.2. $SrFe_{1-x}B_xO_{3-\delta}$

2.2.1. X-ray Diffraction Data

The possible incorporation of borate into $SrFeO_{3-\delta}$ was then examined, with samples of $SrFe_{1-x}B_xO_{3-\delta}$ prepared for $0 \leq x \leq 0.15$. The results showed that a higher borate content was achievable compared to the sulfate-doped samples, with single phase borate-doped samples for $x \leq 0.1$

(Figure 10). In terms of the effect of borate incorporation on the structure, similar results were observed as for the $SrFe_{1-x}S_xO_{3-\delta}$ samples. In particular, borate doping resulted in a similar cubic cell as for the sulfate-doped samples.

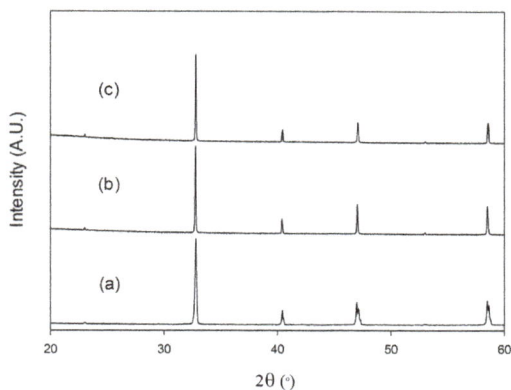

Figure 10. X-ray diffraction patterns for (**a**) $SrFeO_{3-\delta}$, (**b**) $SrFe_{0.95}B_{0.05}O_{3-\delta}$, (**c**) $SrFe_{0.9}B_{0.1}O_{3-\delta}$.

Cell parameters and site occupancies were determined using the Rietveld method (an example fit is shown in Figure 11).

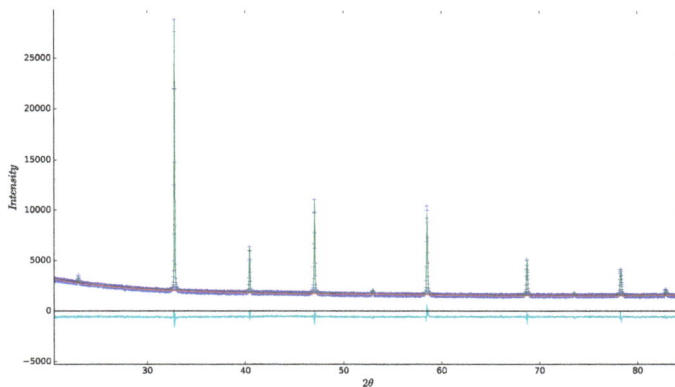

Figure 11. Observed, calculated and difference X-ray diffraction profile for $SrFe_{0.95}B_{0.05}O_{3-\delta}$.

The refinements gave Fe/B occupancies in good agreement with those expected (Table 4). Furthermore, in this case, a decrease in the unit cell was observed on borate doping in agreement with the smaller size of B^{3+} compared to $Fe^{3+/4+}$, and the fact that unlike the situation for sulfate doping there is no additional oxygen associated with the dopant (i.e., we are effectively replacing Fe(III)$O_{1.5}$ with B$O_{1.5}$).

Table 4. Lattice parameters and Fe/B site occupancies obtained from Rietveld analysis using X-ray diffraction data for $SrFe_{1-x}B_xO_{3-\delta}$. The structure of $SrFeO_{3-\delta}$ was refined using a tetragonal space group (P4/mmm). The structures of the doped samples were refined using a cubic space group ($Pm\bar{3}m$).

	$SrFe_{1-x}B_xO_{3-\delta}$		
B (x)	0	0.05	0.1
a (Å)	3.8648(1)	3.8593(1)	3.8561(1)
c (Å)	3.8487(1)	-	-
V (Å^3)	57.486(4)	57.483(2)	57.336(4)
R_{wp} (%)	1.84	3.23	3.67
R_{exp} (%)	0.92	2.30	2.52
Fe occupancy	1	0.92(1)	0.89(1)
B occupancy	-	0.08(1)	0.11(1)
Fe/B Uiso	0.003(1)	0.015(1)	0.022(1)

2.2.2. Stability under N_2

Heat treatment of the $SrFe_{1-x}B_xO_{3-\delta}$ samples at 950 °C under N_2 showed similar results to those observed on sulfate doping, with the borate-containing samples remaining cubic after this heat treatment in nitrogen (Figure 12). In line with the reduction of the Fe oxidation state to Fe^{3+}, there was a shift in the peak positions to lower angles, indicating a larger unit cell.

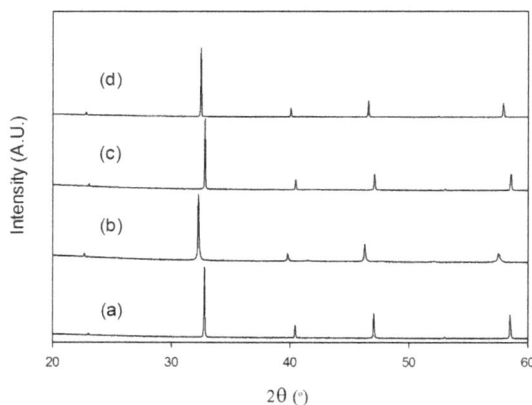

Figure 12. X-ray diffraction patterns for (**a**) $SrFe_{0.95}B_{0.05}O_{3-\delta}$, (**b**) $SrFe_{0.95}B_{0.05}O_{3-\delta}$ after heating under N_2, (**c**) $SrFe_{0.9}B_{0.1}O_{3-\delta}$, (**d**) $SrFe_{0.9}B_{0.1}O_{3-\delta}$ after heating under N_2.

2.2.3. Thermogravimetric Analysis

These samples were then analysed by TGA (heat treatment under N_2). In contrast to the air synthesised $SrFe_{1-x}S_xO_{3-\delta}$ samples, there was no mass loss due to CO_2 observed in these $SrFe_{1-x}B_xO_{3-\delta}$ systems, thus indicating no carbonate present.

2.2.4. Conductivity Data

For the $SrFe_{1-x}B_xO_{3-\delta}$ samples (x = 0.05, 0.1), the conductivity data showed significantly lower conductivities at lower temperatures compared with the undoped system (Figure 13). However, above 600 °C, the conductivities were comparable with the x = 0.1 sample, in particular showing a small improvement in the conductivity compared with undoped $SrFeO_{3-\delta}$. Notably, these temperatures are in the range where operation as a cathode in a solid oxide fuel cell would be.

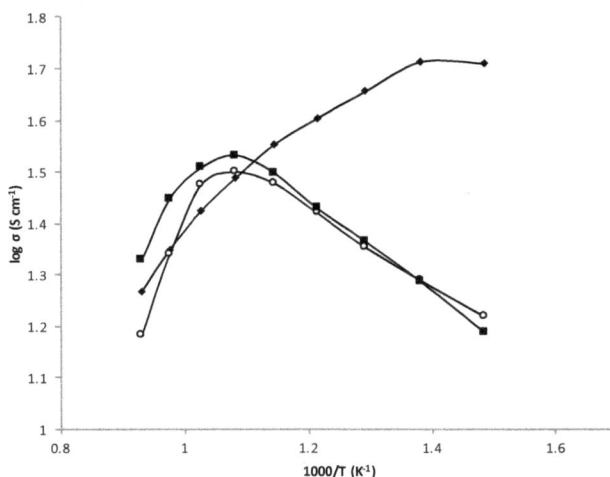

Figure 13. Plot of log σ vs. 1000/T for SrFeO$_{3-\delta}$ (●), SrFe$_{0.95}$B$_{0.05}$O$_{3-\delta}$ (○), SrFe$_{0.9}$B$_{0.1}$O$_{3-\delta}$ (■) in air.

Conductivity values (at 700 °C) for the borate- and sulfate-doped samples are shown in Table 5, and compared to the equivalent data for silicate-doped SrFeO$_{3-\delta}$ (from reference [23]). The data show similar values for all samples at this typical solid oxide fuel cell operating temperature.

Table 5. Conductivity data in air at 700 °C for SrFe$_{1-x}$M$_x$O$_{3-\delta}$ where M = Si [23], S and B.

	Si (x)				S (x)				B (x)	
	0	0.05	0.1	0.15	0	0.025	0.05	0.075	0.05	0.1
Conductivity 700 °C (S cm^{-1})	26	21	35	18	26	25	30	25	30	32

3. Experimental

High-purity SrCO$_3$, Fe$_2$O$_3$, (NH$_4$)$_2$SO$_4$, and H$_3$BO$_3$ were used to prepare SrFe$_{1-x}$S/B$_x$O$_{3-\delta}$ samples. Stoichiometric mixtures of the powders were intimately ground and initially heated to 900 °C (4 °C/min) for 12 h. Samples were then ballmilled (350 rpm for 1 h, Fritsch Pulverisette 7 planetary Mill) and reheated to 1000, 1050 and 1100 °C for 12 h with ballmilling of samples between heat treatments. For the SrFe$_{1-x}$B$_x$O$_{3-\delta}$ (x = 0.05, 0.1) samples, a higher temperature (1200 °C) heat treatment was required to achieve single phase samples. In order to ensure maximum oxygen content, all samples underwent a final heat treatment at 350 °C for 12 h in air.

Additionally, portions of the SrFe$_{1-x}$S$_x$O$_{3-\delta}$ (x = 0, 0.025, 0.05, 0.075) samples were heated to 900 °C under oxygen for 12 h with slow cooling at 50 °C /h to 350 °C, with the samples then maintained at this temperature for 12 h followed by cooling at 50 °C/h to room temperature. To test the stability under low p(O$_2$) conditions, both sulfate and borate-doped samples were heated under N$_2$ to 950 °C for 12 h.

Powder X-ray diffraction data were used in order to determine lattice parameters and phase purity. For SrFe$_{1-x}$S$_x$O$_{3-\delta}$ samples heated in air, X-ray diffraction data were collected on a Panalytical Empyrean diffractometer equipped with a Pixcel 2D detector (Cu Kα radiation). For the remaining SrFe$_{1-x}$S$_x$O$_{3-\delta}$ and SrFe$_{1-x}$B$_x$O$_{3-\delta}$ samples, a Bruker D8 diffractometer with Cu Kα$_1$ radiation was used.

Samples were also analysed using thermogravimetric analysis (Netzch STA 449 F1 Jupiter Thermal Analyser with mass spectrometry attachment). Samples were heated to 950 °C in N_2 (10 °C/min) and held for 30 min to reduce the iron oxidation state to Fe^{3+}.

Pellets for conductivity measurements were prepared as follows: powders of $SrFe_{1-x}S_xO_{3-\delta}$ and $SrFe_{1-x}B_xO_{3-\delta}$ heated in air were initially ball milled (350 rpm for 1 h), before pressing into compacts and sintering at 1100 °C ($SrFe_{1-x}S_xO_{3-\delta}$) and 1200 °C ($SrFe_{1-x}B_xO_{3-\delta}$) for 12 h in air. Four Pt electrodes were attached with Pt paste and the samples were heated at 900 °C for 1 h in air. Samples were then furnace cooled to 350 °C and held at this temperature for 12 h to ensure full oxygenation. Conductivities were measured using the four probe dc method.

4. Conclusions

The results presented in this paper demonstrate the first reports of successful incorporation of sulfate and borate into $SrFeO_{3-\delta}$. This doping strategy results in a change from a tetragonal to a cubic cell, which is maintained even after heating under N_2, where undoped $SrFeO_{3-\delta}$ transforms to the oxygen vacancy ordered brownmillerite structure. Conductivity data in air show that the borate/sulfate-doped samples have comparable conductivities (Table 5) to undoped $SrFeO_{3-\delta}$ at solid oxide fuel cell operating temperatures. Given these initial promising results, further studies are warranted to investigate the performance of these doped systems as solid oxide fuel cell cathodes.

Acknowledgments: We would like to thank EPSRC for funding (studentship for Abbey Jarvis). The raw datasets associated with the results shown in this paper are available from the University of Birmingham archive: http://epapers.bham.ac.uk/3011/.

Author Contributions: P.R.S. conceived and designed the experiments; A.J. performed the experiments; A.J. and P.R.S. analysed the data; A.J. and P.R.S. wrote the paper.

Conflicts of Interest: The authors declare no conflict of interest.

References

1. Orera, A.; Slater, P.R. New Chemical Systems for Solid Oxide Fuel Cells. *Chem. Mater.* **2010**, *22*, 675–690. [CrossRef]
2. Jacobson, A.J. Materials for Solid Oxide Fuel Cells. *Chem. Mater.* **2010**, *22*, 660–674. [CrossRef]
3. Ishihara, T.; Matsuda, H.; Takita, Y. Doped $LaGaO_3$ Perovskite Type Oxide as a New Oxide Ionic Conductor. *J. Am. Chem. Soc.* **1994**, *116*, 3801–3803. [CrossRef]
4. Hancock, C.A.; Porras-Vazquez, J.M.; Keenan, P.J.; Slater, P.R. Oxyanions in Perovskites: From Superconductors to Solid Oxide Fuel Cells. *Dalton Trans.* **2015**, *44*, 10559–10569. [CrossRef] [PubMed]
5. Francesconi, M.G.; Greaves, C. Anion Substitutions and Insertions in Copper Oxide Superconductors. *Supercond. Sci. Technol.* **1997**, *10*, A29–A37. [CrossRef]
6. Slater, P.R.; Greaves, C.; Slaski, M.; Muirhead, C.M. Copper Oxide Superconductors Containing Sulphate and Phosphate Groups. *Phys. C Supercond. Appl.* **1993**, *208*, 193–196. [CrossRef]
7. Letouzé, F.; Martin, C.; Maignan, A.; Michel, C.; Hervieu, M.; Raveau, B. Stabilization of New Superconducting Thallium Cuprates and Oxycarbonates by Molybdenum. *Phys. C Supercond.* **1995**, *254*, 33–43. [CrossRef]
8. Kinoshita, K.; Yamada, T. A New Copper Oxide Superconductor Containing Carbon. *Nature* **1992**, *357*, 313–315. [CrossRef]
9. Huvé, M.; Michel, C.; Maignan, A.; Hervieu, M.; Martin, C.; Raveau, B. A 70 K Superconductor. The Oxycarbonate $Tl_{0.5}Pb_{0.5}Sr_4Cu_2(CO_3)O_7$. *Phys. C Supercond.* **1993**, *205*, 219–224. [CrossRef]
10. Goutenoire, F.; Hervieu, M.; Maignan, A.; Michel, C.; Martin, C.; Raveau, B. A 62 K Superconductor with an Original Structure: $Sr_{4-x}Ba_xTlCu_2CO_3O_7$. *Phys. C Supercond.* **1993**, *210*, 359–366. [CrossRef]
11. Von Schnering, H.G.; Walz, L.; Schwarz, M.; Becker, W.; Hartweg, M.; Popp, T.; Hettich, B.; Müller, P.; Kämpf, G. The Crystal Structures of the Superconducting Oxides $Bi_2(Sr_{1-x}Ca_x)_2CuO_{8-\delta}$ and $Bi_2(Sr_{1-y}Ca_y)_3Cu_2O_{10-\delta}$ with $0 \leq x \leq 0.3$ and $0.16 \leq y \leq 0.33$. *Angew. Chem. Int. Ed. Engl.* **1988**, *27*, 574–576. [CrossRef]

12. Maignan, A.; Pelloquin, D.; Malo, S.; Michel, C.; Hervieu, M.; Raveau, B. Stabilisation of Three New Oxycarbonates by V and Cr Substitutions The Superconductors (Tl,M) $_1Sr_4Cu_2(CO_3)O_7$ (M Cr, V) and (Hg,V) $_1Sr_4Cu_2(CO_3)O_{6+z}$. *Phys. C Supercond.* **1995**, *249*, 220–233. [CrossRef]

13. Shin, J.F.; Orera, A.; Apperley, D.C.; Slater, P.R. Oxyanion Doping Strategies to Enhance the Ionic Conductivity in $Ba_2In_2O_5$. *J. Mater. Chem.* **2011**, *21*, 874–879. [CrossRef]

14. Pérez-Coll, D.; Pérez-Flores, J.C.; Nasani, N.; Slater, P.R.; Fagg, D.P. Exploring the Mixed Transport Properties of sulfur(VI)-Doped $Ba_2In_2O_5$ for Intermediate-Temperature Electrochemical Applications. *J. Mater. Chem. A* **2016**, *4*, 11069–11076. [CrossRef]

15. Shin, J.F.; Hussey, L.; Orera, A.; Slater, P.R. Enhancement of the Conductivity of $Ba_2In_2O_5$ through Phosphate Doping. *Chem. Commun.* **2010**, *46*, 4613–4615. [CrossRef] [PubMed]

16. Shin, J.F.; Apperley, D.C.; Slater, P.R. Silicon Doping in $Ba_2In_2O_5$: Example of a Beneficial Effect of Silicon Incorporation on Oxide Ion/proton Conductivity. *Chem. Mater.* **2010**, *22*, 5945–5948. [CrossRef]

17. Hancock, C.A.; Slade, R.C. T.; Varcoe, J.R.; Slater, P.R. Synthesis, Structure and Conductivity of Sulfate and Phosphate Doped $SrCoO_3$. *J. Solid State Chem.* **2011**, *184*, 2972–2977. [CrossRef]

18. Hancock, C.A.; Slater, P.R. Synthesis of Silicon Doped $SrMO_3$ (M = Mn, Co): Stabilization of the Cubic Perovskite and Enhancement in Conductivity. *Dalton Trans.* **2011**, *40*, 5599–5603. [CrossRef] [PubMed]

19. Liu, Y.; Zhu, X.; Yang, W. Stability of Sulfate Doped $SrCoO_{3-\delta}$ MIEC Membrane. *J. Memb. Sci.* **2016**, *501*, 53–59. [CrossRef]

20. Zhu, Y.; Zhou, W.; Sunarso, J.; Zhong, Y.; Shao, Z. Phosphorus-Doped Perovskite Oxide as Highly Efficient Water Oxidation Electrocatalyst in Alkaline Solution. *Adv. Funct. Mater.* **2016**, *26*, 5862–5872. [CrossRef]

21. Li, M.; Zhou, W.; Xu, X.; Zhu, Z. $SrCo_{0.85}Fe_{0.1}P_{0.05}O_{3-\delta}$ Perovskite as a Cathode for Intermediate-Temperature Solid Oxide Fuel Cells. *J. Mater. Chem. A* **2013**, *1*, 13632–13639. [CrossRef]

22. Porras-Vazquez, J.M.; Kemp, T.F.; Hanna, J.V.; Slater, P.R. Synthesis and Characterisation of Oxyanion-Doped Manganites for Potential Application as SOFC Cathodes. *J. Mater. Chem.* **2012**, *22*, 8287–8293. [CrossRef]

23. Porras-Vazquez, J.M.; Pike, T.; Hancock, C.A.; Marco, J.F.; Berry, F.J.; Slater, P.R. Investigation into the Effect of Si Doping on the Performance of $SrFeO_{3-\delta}$ SOFC Electrode Materials. *J. Mater. Chem. A* **2013**, *1*, 11834–11841. [CrossRef]

24. Starkov, I.; Bychkov, S.; Matvienko, A.; Nemudry, A. Oxygen Release Technique as a Method for the Determination Of "δ-pO2-T" diagrams for MIEC Oxides. *Phys. Chem. Chem. Phys.* **2014**, *16*, 5527–5535. [CrossRef] [PubMed]

© 2017 by the authors. Licensee MDPI, Basel, Switzerland. This article is an open access article distributed under the terms and conditions of the Creative Commons Attribution (CC BY) license (http://creativecommons.org/licenses/by/4.0/).

crystals

MDPI

Article

Nb-Doped $0.8BaTiO_3$-$0.2Bi(Mg_{0.5}Ti_{0.5})O_3$ Ceramics with Stable Dielectric Properties at High Temperature

Feng Si, Bin Tang *, Zixuan Fang and Shuren Zhang

State Key Laboratory of Electronic Thin Films and Integrated Devices,
University of Electronic Science and Technology of China, Chengdu 610054, China;
sifeng928@163.com (F.S.); ZXFANG2015@163.com (Z.F.); zsr@uestc.edu.cn (S.Z.)
* Correspondence: tangbin@uestc.edu.cn

Academic Editor: Haidong Zhou
Received: 6 April 2017; Accepted: 3 June 2017; Published: 11 June 2017

Abstract: Nb-doped $0.8BaTiO_3$-$0.2Bi(Mg_{0.5}Ti_{0.5})O_3$ ceramics were prepared by conventional solid-state method. The dielectric properties and the structural properties were investigated. When Nb_2O_5 is doped into 0.8BT-0.2BMT system, a small amount of $Ba_4Ti_{12}O_{27}$ secondary phase is formed. The lattice parameters gradually increase with the Nb_2O_5 doping. It is found that the temperature-capacitance characteristics greatly depend on Nb_2O_5 content. With the addition of 3.0 mol% Nb_2O_5, a 0.8BT-0.2BMT ceramic sample could satisfy the EIA X9R specification. This material is promising for high-temperature MLCC application.

Keywords: barium titanate; X9R; dielectric ceramics; MLCC

1. Introduction

Multilayer ceramic capacitors (MLCC) have been extensively used in many kinds of electronic products in the last few decades. Some harsh conditions—such as downhole drilling, aerospace, and automotive environment—require the capability of MLCCs to withstand temperatures over 200 °C or more [1,2]. By contrast, the X7R (−55 to 125 °C, $\Delta C/C_{25\ °C} \leq 15\%$) and the X8R (−55 to 150 °C, $\Delta C/C_{25\ °C} \leq 15\%$)-type MLCCs defined by the Electronic Industries Association (EIA) standards are not competent because their ceiling temperatures are 125 °C and 150 °C, respectively [3,4]. Therefore, it is of great significance to exploit a material used for X9R (−55 to 200 °C, $\Delta C/C_{25\ °C} \leq 15\%$)-type MLCCs, and develop a new temperature-stable dielectric material with high-temperature resistance, high dielectric constant, and low dielectric loss.

As well known, high permittivity MLCCs are generally derived from perovskite structures, such as barium titanate (BT) dielectric ceramics. As the Curie temperature (T_c) of BT is approximately 130 °C, however, it is difficult to prepare pure $BaTiO_3$ (BT) ceramics with weak temperature dependence [5]. Recently, $Bi(Me)O_3$ based perovskite (Me = Sc, Al, $Zn_{1/2}Ti_{1/2}$, etc.) were of interest for the study and development of new dielectric ceramics because of their relatively high dielectric constant, broad diffused dielectric behavior, and high Curie temperature [6–9]. Therefore, perovskite solid solutions based on BT-$Bi(Me)O_3$ composition with a high permittivity are developed and explored for high temperature capacitor applications [10]. The 0.8BT-0.2BMT ceramic with moderate dielectric constant and dielectric loss is reported to possess lower temperature variation of capacitance in high temperature regions [11]. The Nb_2O_5 doped BT systems are reported for potential application in advanced X8R capacitors, but the Nb_2O_5 dopant depressed dielectric constant in exchange for temperature stability [12]. The Nb^{5+} ion was found to diffuse into the crystal lattice and form the chemically inhomogenous regions, called 'core-shell' structure, showing good dielectric stability [13–15]. In this study, the 0.8BT-0.2BMT dielectric system is modified only by Nb_2O_5 to improve the dielectric temperature stability.

2. Experimental Procedure

Nb-doped $0.8BaTiO_3$-$0.2Bi(Mg_{0.5}Ti_{0.5})O_3$($0.8BT$-$0.2BMT$) ceramics were prepared using solid-state reaction method. $BaTiO_3$ (99.9%), Bi_2O_3 (99.0%), MgO (99%), and TiO_2 (99.0%) powders and Nb_2O_5 (99.5%) with different doped levels ($0 \leq x \leq 3$ mol%) were batched stoichiometrically and ball milled in a polyethylene jar with yttria-stabilized zirconia balls and de-ionized water for 4 h at a rate of 280 rpm. Then the mixtures were dried and calcined in alumina crucibles at 900 °C for 2 h. Subsequently, the calcined powders were then re-milled in de-ionized water using zirconia balls for 4 h. After drying, the resulting powders were granulated with a few drops of PVA binder and pressed into disk with 12 mm in diameter and 1 mm in thickness. Samples were sintered at 1120 °C for 2 h in air with a heating rate of 3 °C/min and then cooling with furnace to ambient temperature.

Crystal structure of the samples was determined at room temperature by X-ray powder diffraction (XRD) (Cu Kα radiation, PANalytical X'Pert Pro, Almelo, The Netherlands). Rietveld refinement of the crystal structures was performed using the GSAS-EXPGUI program [16]. Microstructure and composition analysis of the ceramic pellets were performed by scanning electron microscopy (SEM, Inspect F, FEI Company, Eindhoven, The Netherlands), which was operated at 20 kV. After samples were fired at 800 °C with silver paste on both sides, the room temperature dielectric constant and dielectric loss were measured by use of a precision LCR meter (4284A, Aglient) at 1 kHz and 1 Vrms. The temperature-capacitance characteristics of the samples were measured from −55 °C to 200 °C with LCR meter (4284A, Aglient) and an automatic temperature controller at 1 kHz and 1 Vrms with a rate of 2 °C/min. The insulation resistance of the specimens was recorded by a Digital Super Megohm Meters (DSM-8104, HIOKI, Nagano, Japan) at 25 °C. The bulk density was measured using the Archimedes method.

3. Results and Discussion

Figure 1 shows the XRD patterns of well-sintered Nb_2O_5 doped 0.8BT-0.2BMTceramic disks. All the samples presented a desired perovskite structure and no secondary phase was discovered for sample without Nb_2O_5 dopants. It indicated that homogeneous solid solution of 0.8BT-0.2BMT was obtained. However, a small amount of secondary phase $Ba_4Ti_{12}O_{27}$ is found in all Nb-doped 0.8BT-0.2BMT samples. The phase structures are analyzed by General Structure Analysis System (GSAS) Rietveld refinements for 0.8BT-0.2BMT with various amount of Nb_2O_5, as shown in Figure 2. The main phase is detected as $BaTiO_3$ (P4mm, ICSD No. 086286). The lattice parameters and reliability factors are listed in Table 1. All the lattice parameters are obtained with high reliability as $R_{wp} < 10\%$, $R_p < 8\%$. The lattice parameters calculated from the Rietveld refinements increase continuously with the increase of Nb_2O_5 content. This indicated the larger ionic radius substitution that occurred according to Bragg's law. In other words, Nb^{5+} mainly enters the Ti-sites as a donor because the Nb^{5+} (0.64 Å) is a considerably larger radius than that of Ti^{4+} (0.605 Å) in six-fold coordination [17]. Meanwhile, Nb substitution on Ti site in the $BaTiO_3$ lattice will leave the excess Ti out of the grains, which is responsible for secondary phase formation at the grain boundary.

Table 1. Lattice parameters from Rietveld refinement for the samples.

Samples	Lattice Parameters			Reliability Factors		
	a (Å)	c (Å)	V (Å^3)	R_{wp}	R_p	χ^2
+0.0 mol% Nb_2O_5	4.0096	4.0131	64.5182	0.0786	0.0598	1.678
+1.0 mol% Nb_2O_5	4.0112	4.0140	64.5842	0.0873	0.0651	2.187
+2.0 mol% Nb_2O_5	4.0131	4.0165	64.6856	0.0780	0.0579	1.555
+3.0 mol% Nb_2O_5	4.0150	4.0179	64.7695	0.0847	0.0621	1.943

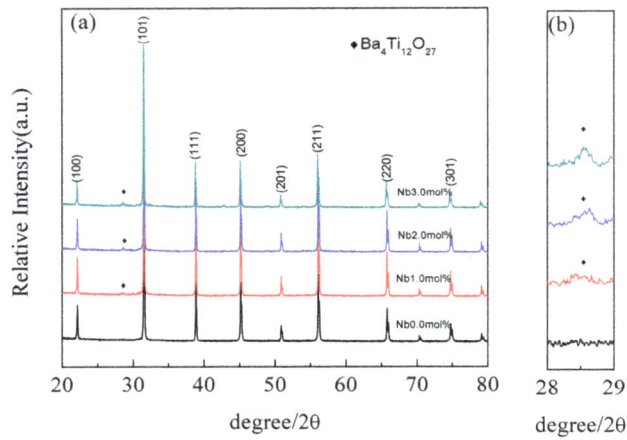

Figure 1. (**a**) The XRD patterns of well-sintered Nb_2O_5 doped 0.8BT-0.2BMT ceramic disks; (**b**) the XRD patterns in the region between 28 and 29 degree.

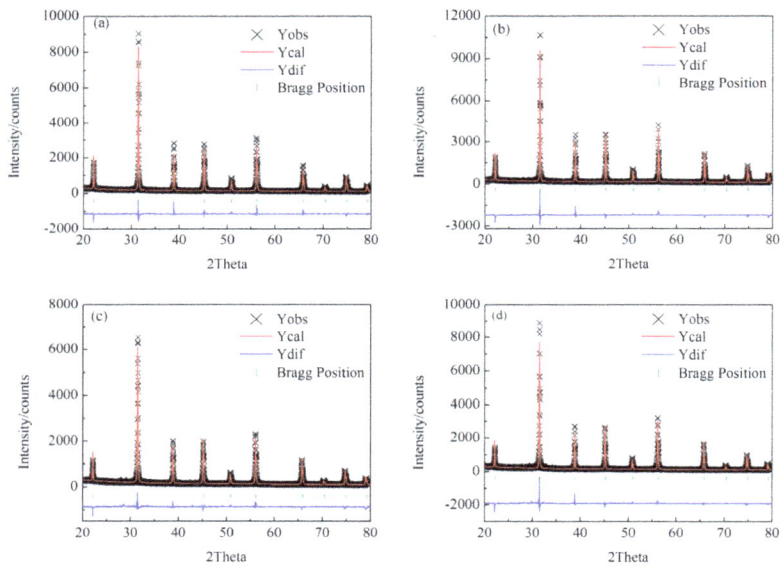

Figure 2. XRD profiles for the Rietveld refinement resultsfor 0.8BT-0.2BMTwith various amount of Nb_2O_5: (**a**) 0.0 mol%; (**b**) 1.0 mol%; (**c**) 2.0 mol%; (**d**) 3.0 mol%.

Figure 3 shows the microstructure of various amount of Nb_2O_5 doped 0.8BT-0.2BMT ceramics. It is observed that all the Nb-doped 0.8-0.2BZT ceramics are in high densification with homogeneous grain size being on the order of several micrometers. There is a small change of the grain size as the Nb_2O_5 content increases, where the average grain size is about 1.5 μm. Some stick grains, pointed by arrows in the Nb_2O_5 doped samples, are considered as secondary phase ($Ba_4Ti_{12}O_{27}$). This result is highly consistent with XRD analysis.

Figure 3. The microstructure of various amount of Nb_2O_5 doped 0.8BT-0.2BMT ceramics: (**a**) 0.0 mol%; (**b**) 1.0 mol%; (**c**) 2.0 mol %; (**d**) 3.0 mol%.

Main dielectric and electrical properties at room temperature of the samples are listed in Table 2. As we can see, dielectric constant at 25 °C decreased with increasing Nb_2O_5 content. In the Nb_2O_5 modified BT-BMT system, the Nb^{5+} will diffuse into the crystal lattice to form the chemically inhomogeneous structure ('core-shell'). The dielectric characteristics of the barium titanate-based core-shell grain structure were a superposition of the ferroelectric grain core, paraelectric grain shell [18]. With further addition of Nb_2O_5 dopants, the volume fractions of the shell region increased, leading to the decrease of the dielectric constant.

Table 2. Dielectric and electrical properties of the samples at room temperature.

Sample	ε_r	tan δ	$\Delta C/C_{25\,°C} < \pm 15\%$	Resistivity (Ω·cm)	Density (g/cm^3)
+0.0 mol% Nb_2O_5	1683	5.0%	-	1.0×10^{13}	6.214
+1.0 mol% Nb_2O_5	1247	1.5%	−26~197 °C	6.4×10^{11}	6.219
+2.0 mol% Nb_2O_5	925	0.7%	−40~200 °C	5.9×10^{12}	6.148
+3.0 mol% Nb_2O_5	764	0.5%	−55~200 °C	1.3×10^{13}	6.159

When a small amount of Nb_2O_5 is doped into the BT matrix, electronic compensation regime appears [19–21].

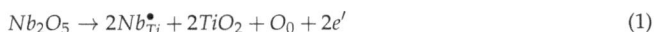

$$Nb_2O_5 \rightarrow 2Nb^{\bullet}_{Ti} + 2TiO_2 + O_0 + 2e' \tag{1}$$

The insulation resistivity of the sample with 0.1 mol% Nb_2O_5 dopants decreased for the electronic compensation. With the increasing of Nb_2O_5 concentration, the electronic compensation changes into Ti vacancy compensation.

$$2Nb_2O_5 \rightarrow 4Nb_{Ti}^{\bullet} + 5TiO_2 + V_{Ti}'''' \tag{2}$$

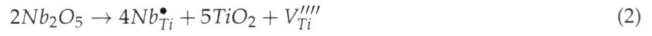

The mechanism is mainly involved with the compensation of Ti vacancies so that the samples with 3.0 mol% Nb_2O_5 can become highly insulating as shown in Table 2 [22].

Figure 4 demonstrates the temperature dependence of dielectric constant and dielectric loss for 0.8BT-0.2BMT with various amount of Nb_2O_5. Double dielectric peaks are observed: one at around 0 °C and a second, smaller, peak at about 130 °C. The curves are flattened by the Nb_2O_5 dopants at the cost of deteriorating dielectric constant. However, when the doped level is above 2.0 mol%, it is difficult to detect the higher permittivity peak. Loss tangent data at room temperature are lower than 5%, and exhibit a decrease with increasing Nb_2O_5 content. Dielectric loss at low temperature is rather high and decreases monotonously with the increasing temperature.

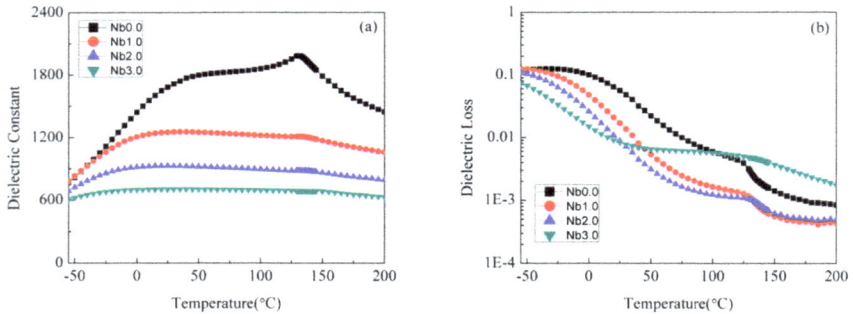

Figure 4. Temperature dependence of (**a**) dielectric constantand (**b**) dielectric loss for 0.8BT-0.2BMT with various amount of Nb_2O_5.

Temperature dependence of capacitance variation rate based on $C_{25\,°C}$ for 0.8BT-0.2BMT with various amount of Nb_2O_5 is shown in Figure 5. It can be seen that the temperature-capacitance characteristics have a remarkable improvement with the incorporation of Nb_2O_5, especially at the cold end. At the hot end, the permittivity peak is suppressed with the increase of Nb_2O_5 dopants. The 3.0 mol% Nb_2O_5-doped 0.8BT-0.2BMT ceramic sample is found to satisfy the EIA X9R specification with a permittivity of 764 and dielectric loss of 0.5%. The existence of the double dielectric anomalies over the studied temperature range, which is the characteristic phenomenon of 'core–shell' structure in BT ceramics, is beneficial to improving the temperature stability of the dielectric properties [20]. The Nb^{5+} will diffuse into the crystal lattice to form the chemically inhomogeneous structure, where the core is pure BT, while the Nb ion mainly concentrated in the shell region. In the ceramics with 'core–shell' structure, the double peaks in the temperature dependence of dielectric constant curve can usually be obtained [14]. In the Nb_2O_5 modified 0.8BT-0.2BMT system, the dopant oxides tend to accumulate at the grain boundary and diffuse simultaneously into BT grain. Nevertheless, the diffusion of Bi ions was weakened due to the slow diffusion of Nb^{5+} during the sintering process. Thus Bi_2O_3, MgO, and TiO_2 mainly stayed in the shell region which should be responsible for the lower dielectric peak at ~0 °C, while pure BT at cores is attributable to the dielectric peak at ~130 °C, as shown in Figure 4.

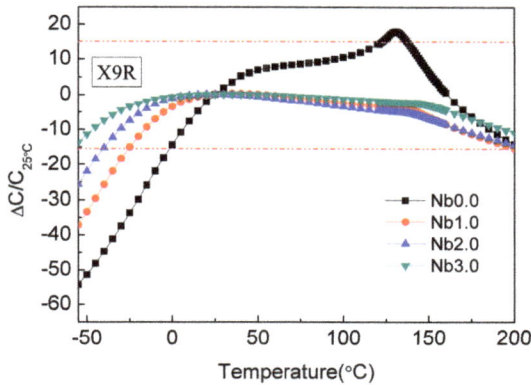

Figure 5. Temperature dependence of capacitance variation rate based on $C_{25\,°C}$ for 0.8BT-0.2BMT with various amount of Nb_2O_5.

4. Conclusions

In this work, we investigated the effect of Nb_2O_5 doping on the dielectric properties of 0.8BT-0.2BMT ceramics. A small amount of $Ba_4Ti_{12}O_{27}$ secondary phase is observed when Nb_2O_5 doped into 0.8BT-0.2BMT system. The dielectric temperature stability is strongly affected by the Nb_2O_5 dopants. The 3.0 mol% Nb_2O_5-doped 0.8BT-0.2BMT ceramic sample satisfies the requirement of EIA X9R specification with a permittivity of 764. Moreover, it possesses low dielectric loss and high insulation resistance, which ensures the application of this material.

Acknowledgments: This work is supported by the Open Foundation of National Engineering Research Center of Electromagnetic Radiation Control Materials (ZYGX2016K003-5) and by National Natural Science Foundation of China (Grant No. 51402039)

Author Contributions: Bin Tang, Shuren Zhang and Feng Si conceived and designed the experiments; Feng Si performed the experiments; Bin Tang and Feng Si analyzed the data; Zixuan Fang contributed reagents/materials/analysis tools; Feng Si wrote the paper.

Conflicts of Interest: The authors declare no conflict of interest.

References

1. Johnson, R.W.; Evans, J.L.; Jacobsen, P. The changing automotive environment: High-temperature electronics. *IEEE Trans. Electron. Pack. Manuf.* **2004**, *27*, 164–176. [CrossRef]

2. Werner, M.R.; Fahrner, W.R. Review on materials, microsensors, systems, and devices for high-temperature and harsh-environment applications. *IEEE Trans. Ind. Electron.* **2001**, *48*, 249–257. [CrossRef]

3. Lee, W.H.; Su, C.Y. Improvement in the temperature stability of a $BaTiO_3$-Based multilayer ceramic capacitor by constrained sintering. *J. Am. Ceram. Soc.* **2007**, *90*, 3345–3348. [CrossRef]

4. Wang, S.F.; Dayton, G.O. Dielectric properties of fine-grained barium titanate based X7R materials. *J. Am. Ceram. Soc.* **1999**, *82*, 2677–2682. [CrossRef]

5. Sakayori, K.; Matsui, Y.; Abe, H. Curie-Temperature of Batio3. *Jpn. J. Appl. Phys.* **1995**, *34*, 5443–5445. [CrossRef]

6. Datta, K.; Thomas, P.A. Structural investigation of a novel perovskite-based lead-free ceramics: $xBiScO_3$-(1-x)$BaTiO_3$. *J. Appl. Phys.* **2010**, *107*, 043516. [CrossRef]

7. Raengthon, N.; Cann, D.P. Dielectric Relaxation in $BaTiO_3$-Bi($Zn_{1/2}Ti_{1/2}$)O_3 Ceramics. *J. Am. Ceram. Soc.* **2012**, *95*, 1604–1612. [CrossRef]

8. Liu, M.Y.; Hao, H.; Zhen, Y.C. Temperature stability of dielectric properties for $xBiAlO_3$-(1-x)$BaTiO_3$ ceramics. *J. Eur. Ceram. Soc.* **2015**, *35*, 2303–2311. [CrossRef]

9.   Wang, Y.R.; Pu, Y.P.; Zheng, H.Y. Enhanced dielectric relaxation in (1-x)BaTiO$_3$-xBiYO$_3$ ceramics. *Mater. Lett.* **2016**, *181*, 358–361. [CrossRef]

10. Zeb, A.; Milne, S.J. High temperature dielectric ceramics: A review of temperature-stable high-permittivity perovskites. *J. Mater. Sci. Mater. Electron.* **2015**, *26*, 9243–9255. [CrossRef]

11. Ren, P.R.; Wang, X.; Fan, H.Q. Structure, relaxation behaviors and nonlinear dielectric properties of BaTiO$_3$-Bi(Ti$_{0.5}$Mg$_{0.5}$)O$_3$ ceramics. *Ceram. Int.* **2015**, *41*, 7693–7697. [CrossRef]

12. Sun, Y.; Liu, H.X.; Hao, H. Structure Property Relationship in BaTiO$_3$-Na$_{0.5}$Bi$_{0.5}$TiO$_3$-Nb$_2$O$_5$-NiO X8R System. *J. Am. Ceram. Soc.* **2015**, *98*, 1574–1579. [CrossRef]

13. Hennings, D.F.K. Dielectric materials for sintering in reducing atmospheres. *J. Eur. Ceram. Soc.* **2001**, *21*, 1637–1642. [CrossRef]

14. Chazono, H.; Kishi, H. Sintering characteristics in the BaTiO$_3$-Nb$_2$O$_5$-Co$_3$O$_4$ ternary system: II, stability of so-called "core–shell" structure. *J. Am. Ceram. Soc.* **2000**, *83*, 101–106. [CrossRef]

15. Liu, X.A.; Cheng, S.G.; Randall, C.A. The core–shell structure in ultrafine X7R dielectric ceramics. *J. Korean Phys. Soc.* **1998**, *32*, S312–S315.

16. Toby, B.H. EXPGUI, a graphical user interface for GSAS. *J. Appl. Crystallogr.* **2001**, *34*, 210–213. [CrossRef]

17. Wang, T.; Jin, L.; Li, C.C. Relaxor Ferroelectric BaTiO$_3$-Bi(Mg$_{2/3}$Nb$_{1/3}$)O$_3$ Ceramics for Energy Storage Application. *J. Am. Ceram. Soc.* **2015**, *98*, 559–566. [CrossRef]

18. Park, Y.; Kim, H.G. Dielectric temperature characteristics of cerium-modified barium titanate based ceramics with core–shell grain structure. *J. Am. Ceram. Soc.* **1997**, *80*, 106–112. [CrossRef]

19. Brzozowski, E.; Castro, M.S.; Foschini, C.R. Secondary phases in Nb-doped BaTiO$_3$ ceramics. *Ceram. Int.* **2002**, *28*, 773–777. [CrossRef]

20. Kowalski, K.; Ijjaali, M.; Bak, T. Electrical properties of Nb-doped BaTiO$_3$. *J. Phys. Chem. Solid.* **2001**, *62*, 543–551. [CrossRef]

21. Li, W.; Qi, J.Q.; Wang, Y.L. Doping behaviors of Nb$_2$O$_5$ and Co$_2$O$_3$ in temperature stable BaTiO$_3$-based ceramics. *Mater. Lett.* **2002**, *57*, 1–5. [CrossRef]

22. Smyth, D.M. The defect chemistry of donor-doped BaTiO$_3$: A rebuttal. *J. Electroceram.* **2002**, *9*, 179–186. [CrossRef]

© 2017 by the authors. Licensee MDPI, Basel, Switzerland. This article is an open access article distributed under the terms and conditions of the Creative Commons Attribution (CC BY) license (http://creativecommons.org/licenses/by/4.0/).

crystals

MDPI

Article

Microstructure and Electrical Properties of Fe,Cu Substituted (Co,Mn)$_3$O$_4$ Thin Films

Dagmara Szymczewska [1], Sebastian Molin [2,*], Peter Vang Hendriksen [2] and Piotr Jasiński [1]

[1] Faculty of Electronics, Telecommunications and Informatics, Gdansk University of Technology, ul. G. Narutowicza 11/12, 80-233 Gdansk, Poland; dagmara.szymczewska@pg.gda.pl (D.S.); pijas@eti.pg.gpa.pl (P.J.)

[2] Department of Energy Conversion and Storage, Technical University of Denmark, Risø Campus, Frederiksborgvej 399, 4000 Roskilde, Denmark; pvhe@dtu.dk

* Correspondence: sebmo@dtu.dk; Tel.: +45-46-775-296

Academic Editor: Stevin Pramana
Received: 19 May 2017; Accepted: 17 June 2017; Published: 23 June 2017

Abstract: In this work, thin films (~1000 nm) of a pure MnCo$_2$O$_4$ spinel together with its partially substituted derivatives (MnCo$_{1.6}$Cu$_{0.2}$Fe$_{0.2}$O$_4$, MnCo$_{1.6}$Cu$_{0.4}$O$_4$, MnCo$_{1.6}$Fe$_{0.4}$O$_4$) were prepared by spray pyrolysis and were evaluated for electrical conductivity. Doping by Cu increases the electrical conductivity, whereas doping by Fe decreases the conductivity. For Cu containing samples, rapid grain growth occurs and these samples develop cracks due to a potentially too high thermal expansion coefficient mismatch to the support. Samples doped with both Cu and Fe show high electrical conductivity, normal grain growth and no cracks. By co-doping the Mn, Co spinel with both Cu and Fe, its properties can be tailored to reach a desired thermal expansion coefficient/electrical conductivity value.

Keywords: manganese cobalt spinel; high temperature protective coatings; thin films; electrical conductivity

1. Introduction

Ceramic materials are important in engineering. Many future efficient energy conversion and storage technologies depend on the development of new electroceramic materials with desired and tailored properties [1–5]. Materials based on MnCo$_2$O$_4$ are interesting for a broad range of applications, from room temperature to high temperatures (800 °C). They can be used in Li-ion battery electrodes [6,7], electrochemical supercapacitors [8] and as coating materials for steel interconnects for fuel cells [9–11]. The microstructure and the composition of the spinel strongly affect its performance and applicability in a specific technology. In recent years, doping of the spinel by either Fe or Cu has been pursued in order to increase its electrical conductivity, enhance sintering and possibly alter the thermal expansion coefficient [12,13]. Only very recently has simultaneous substitution by both Fe and Cu been reported (on bulk samples) [13].

Manganese cobalt spinel is a preferred material for high temperature protective coatings for steel interconnects for Solid Oxide Fuel/Electrolysis stacks [14]. Its high electrical conductivity and the low mobility of chromium in the material are two important qualifying factors [15]. For high temperature protective coatings, thicknesses > 10 μm are typically used, whereas in applications like supercapacitors and Li-ion batteries, thin films or composites are typically used.

Electrical conductivity of thin films of several materials showed interesting behavior. Effects like ionic conductivity and activation energy dependent on the grain size have been reported [16]. For example, Rupp et al. have studied an ionically (O^{2-}) conducting CGO (cerium-gadolinium oxide) and YSZ (yttria-stabilized zirconia) thin films deposited by spray pyrolysis [17,18]. These studies have revealed very interesting properties of thin films prepared by the low temperature spray pyrolysis process.

Spray pyrolysis is a solution based deposition method, in which liquids are deposited on a heated substrate via droplets from a pressurized atomizer. In comparison to other solution based deposition methods, e.g., spin coating, thermal decomposition of the liquid precursor occur directly on the surface during the deposition process [19–22]. Therefore, spray pyrolysis is a cost and time effective method, well suited for fabrication of thin nanocrystalline films [19]. An important advantage of the solution based methods is their easy ability to modify the chemical composition of the desired products by adding dopants to the liquid precursor.

In this work, $MnCo_2O_4$ spinel and Cu/Fe substituted derivatives ($MnCo_{1.6}Cu_{0.4}O_4$, $MnCo_{1.6}Fe_{0.4}O_4$, $MnCo_{1.6}Cu_{0.2}Fe_{0.2}O_4$) are deposited in the form of thin (~1 μm) films by spray pyrolysis. Sapphire was used as a substrate and the in-plane electrical properties and microstructure of the films were characterized as a function of temperature up to 800 °C.

2. Materials and Methods

Four kinds of polymeric precursor solutions were prepared in order to create a dense layer on a 500 μm thick c-plane sapphire substrate by the spray pyrolysis method. The concentration of cations in each precursor was fixed at 0.2 mol/L. For precursor preparation, $Mn(NO_3)_2 \cdot 4H_2O$, $Co(NO_3)_2 \cdot 6H_2O$, $Cu(NO_3)_2 \cdot 3H_2O$ and $Fe(NO_3)_3 \cdot 9H_2O$ nitrate salts were used. For $MnCo_2O_4$, $MnCo_{1.6}Cu_{0.4}O_4$, $MnCo_{1.6}Fe_{0.4}O_4$ and $MnCo_{1.6}Cu_{0.2}Fe_{0.2}O_4$ nitrate salts were mixed in a stoichiometric ratio 1:2, 1:1.6:0.4, 1:1.6:0.4 and 1:1.6:0.2:0.2, respectively. Salts were dissolved in a mixture of deionized water, diethylene glycol and tetraethylene glycol (1:1:8 vol % respectively). For spray pyrolysis, the following deposition parameters were used: hot plate temperature −390 °C, precursor flow rate −1 mL/h, nozzle-surface distance −600 mm, air pressure −2 bars. For the desired 1 μm layer thickness, 20 mL of the precursor was used. For each composition, two samples with an approximate dimension of 1×1 cm^2 were prepared. One has been used for the electrical conductivity study, the second for the in-situ high temperature c.

The deposited layers were characterized for their electrical conductivity using the van der Pauw method. To get more information about crystallization and electrical conductivity, five heating-cooling cycles were followed, as presented in Figure 1. In each cycle, the samples were heated with a rate of 100 °C/h until the maximum designed temperature was reached. Every cycle had different maximum temperatures starting with 400 °C for the 1st cycle, and 500 °C, 600 °C, 700 °C and 800 °C for the subsequent cycles. For each cycle, the temperature was held at maximum for 1 h. The data were collected from the highest temperature of the cycle and at hold during cooling down to 200 °C at every 25 °C.

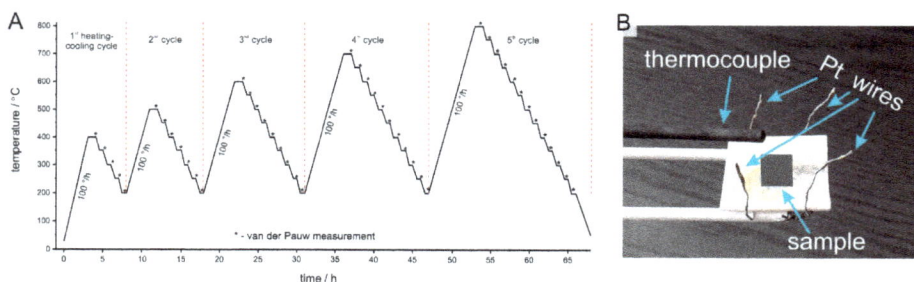

Figure 1. Schematic of the temperature profile (**A**) and sample holder for van der Pauw measurements (**B**).

For microstructural analysis of the layers, scanning electron microscope (SEM) with energy dispersive spectroscopy (EDS) and X-ray diffractometry (XRD) were used. A Hitachi TM3000 with

Bruker Quantax 70 and a Zeiss Supra 35 FEG-SEM were used for characterization of both surfaces and fracture cross-sections. For XRD analysis, a Bruker D8 Advance with CuKα radiation was used. Characterization in both a standard 2θ configuration and in a grazing incidence mode were carried out. In-situ high temperature XRD measurements were carried out using an MRI heating stage based on a Pt-Rh heating element. XRD patterns were used for phase identification, determination of the crystallite size and crystal structure.

The crystallite size was calculated according to the Scherrer formula and the influence of temperature was evaluated by the procedure presented in [18]. The following equation has been used:

$$d = \frac{K\lambda}{\beta \cos \theta} \tag{1}$$

where: The symbols have the following meaning: d—crystallite size, K—shape factor, taken as 0.94, λ—X-ray wavelength, β—broadening at FWHM, θ—Bragg angle.

As the coatings were very thin and deposited on a sapphire substrate, Rietveld refining of the unit cell was not successful. Instead, for qualitative description, a simple approach of calculation of the lattice parameter of the cubic crystal lattice was used. Based on the position of the observed diffraction peaks in the XRD pattern, the lattice parameter was calculated from [18].

$$a = \frac{\lambda \sqrt{h^2 + k^2 + l^2}}{2 \sin \theta} \tag{2}$$

where: A—lattice parameter (Å); h, k, l—the Miller indices of the considered Bragg reflection θ, λ—X-ray wavelength (Å).

3. Results and Discussion

3.1. Analysis of the Produced Coatings

Scanning electron microscopy images of the as-prepared $MnCo_{1.6}Fe_{0.4}O_4$ coatings are shown in Figure 2A,B at two different magnifications (500× and 10,000×). The surfaces of other coatings looked very similar. Due to the low processing temperature of the deposition process (~380 °C), no characteristic grain features are visible by the SEM. The coatings are continuous, they cover the surface well and no cracks nor defects are detected. The chemical compositions of the coatings have been evaluated by the EDS analysis at a low magnification (500×) to average over a large area. The EDS spectra show a presence of Mn, Co and the respective dopants, Cu and/or Fe, as illustrated in Figure 2C–F. Calculated compositions are summarized in Table 1. XRD patterns of the as-produced coatings are shown in Figure 2G. Measurements have been performed in a grazing incidence (GI) mode at an x-ray source angle of 1.5° to avoid a strong signal from the sapphire substrate. Diffraction patterns reveal broad peaks due to small crystallites and especially for the pure $MnCo_2O_4$ show possible presence of some amorphous phase (broad peak around 28°). Even at this relatively low processing temperature of ~380 °C, a cubic spinel phase has crystallized. For reference, positions of the peaks according to JCPDS-ICDD card number 23-1237 are also plotted in Figure 2G. No preferred orientation of the as-deposited film can be detected, as the intensities of the respective peaks seem to resemble the reference pattern.

The chemical composition of the coatings (determined from the EDS analysis), given in Table 1, agrees with the desired stoichiometry, though a slightly higher B atom content (in an AB_2O_4 structure nomenclature) is observed (~2.2 instead of the expected 2). This difference from the desired stoichiometry, noticed in all samples, might be due to a different cation content in the used nitrate salt source. Higher Co content can be tolerated as it should help to stabilize the cubic phase, as phases with a lower Co content (<2) can separate into a mixture of tetragonal Mn_2CoO_4 and cubic $MnCo_2O_4$ spinel [23]. Based on the cation content and assuming (based on the XRD) the existence of only the pure spinel phase, the experimentally determined stoichiometry of the spinel has been calculated.

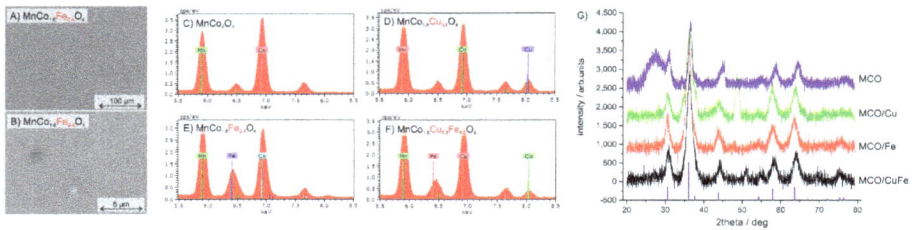

Figure 2. SEM surface image of the MCO/Fe layer (**A,B**) with EDS spectra of all prepared layers (**C–F**) for compositional analysis and XRD patterns (**G**) of prepared coatings (grazing incidence at 1.5 deg).

Table 1. Chemical composition of coatings determined by energy dispersive spectroscopy (EDS) analysis.

	$MnCo_2O_4$	$MnCo_{1.6}Cu_{0.4}O_4$	$MnCo_{1.6}Fe_{0.4}O_4$	$MnCo_{1.6}Cu_{0.2}Fe_{0.2}O_4$
	MCO	MCO/Cu	MCO/Fe	MCO/CuFe
O	60.0	56.7	60.2	58.5
Mn	12.3	13.4	12.7	12.6
Co	27.7	23.9	21.6	22.9
Cu	0	6.0	0	3.2
Fe	0	0	5.5	2.8
(Co + Fe + Cu)/Mn	2.25	2.23	2.14	2.30
EDS determined stoichiometry	$Mn_{0.92}Co_{2.08}$	$Mn_{0.93}Co_{1.66}Cu_{0.42}$	$Mn_{0.96}Co_{1.63}Fe_{0.41}$	$Mn_{0.91}Co_{1.66}Cu_{0.23}Fe_{0.20}$

3.2. Electrical Characterization

The as-produced coatings have been characterized electrically following the temperature profile shown in Figure 1. With each consecutive measurement, the samples have been held at maximum temperature for 1 h and then the electrical conductivity has been measured during cooling. Measurement/temperature profile has been selected to check the possible influence of crystallization, crystallite/grain growth on electrical conductivity. Recorded conductivity values are presented in an Arrhenius type plots for the four samples in Figure 3A–D.

With an initial increase in the temperature from 400 °C (~deposition temperature) to 500 °C and to 600 °C, conductivity increases for all samples. The least increase was observed for the unmodified spinel, where electrical conductivity values do not change for exposures of 400 °C and 500 °C.

Between 500 °C and 400 °C, a change in slope of the conductivity curves is observed. These effects have been reported previously for similar spinel compositions [11,24]. One possible explanation is a change in the prevailing conduction mechanism. Typically, at low temperatures, grain boundaries are considered more conductive than grains and at high temperatures, grains contribute more.

It is interesting to note, that for the exposure temperatures higher than 700 °C, a drop in electrical conductivity occurs. This happens to MCO, MCO/Fe and MCO/CuFe samples. So the maximum electrical conductivity is achieved for layers processed at ~700 °C.

Increase and change of the electrical conductivity of the layers can be due to several factors. The deposition of the layers has been performed at ~390 °C and subsequently, the samples have been evaluated by XRD and measured electrically. As evidenced by the XRD, even at this low temperature, a 1crystalline phase is obtained for all samples. Clearly, for the pure spinel, some amorphous phase is still present. This might be possible also for the other samples, though this is not seen in the XRD spectra.

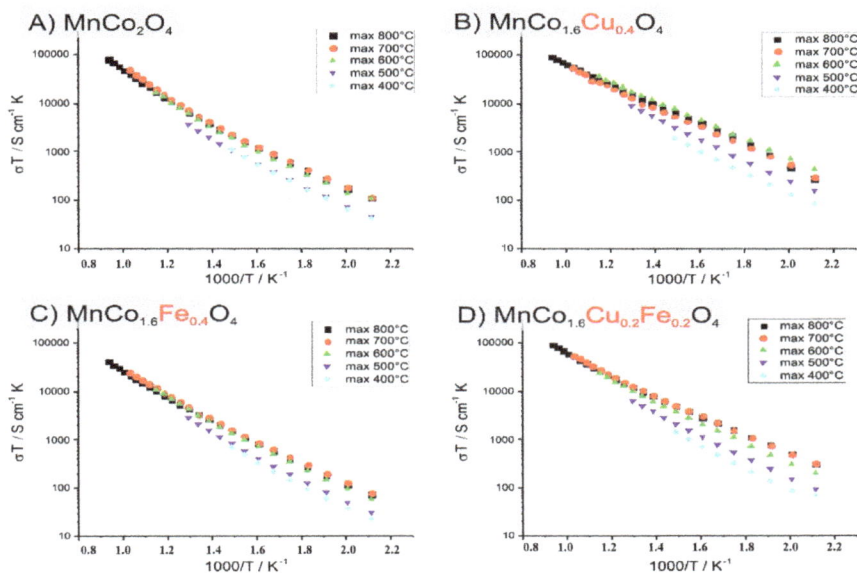

Figure 3. (A–D) Electrical conductivity as a function of temperature plots for all samples after different heat treatments.

One possibility that may account for the increase of conductivity is an increase in the degree of crystallization of the layer. The amount of the well conducting crystalline phase increases with temperature (assuming that the amorphous phase would have a lower/negligible conductivity). However, as the slope of the electrical conductivity (activation energy) curves do not change much for the samples processed at both high and low temperatures, it might be assumed that the crystalline phase (grains plus grain boundaries) has been already continuous at low temperatures. Therefore, an observed 3–4 fold increase in the electrical conductivity cannot be simply explained by a change in the ratio of the crystalline to the amorphous phase. Therefore, the main mechanism responsible for the changes of the total measured electronic conductivities might be the change of the size and thus the number of crystallites, grains and grain boundaries. The total electronic conductivity is determined by both the grain and grain boundaries contribution (σ_G and σ_{GB}), with typically different thermal activation. For the low exposure temperatures, with small grains, the ratio of the number (or volume part) of grain boundaries to grains would be high and with increasing temperature it decreases. This seems to be the dominating mechanism and will be studied further in future works.

For the MCO/Cu layer, some irregularities in the curves are observed, especially during the measurement with the maximum temperature of 700 °C. As will be described in more detail later, this layer had cracked and thus these results should be taken with caution.

As already mentioned, a complex behavior of the evolution of the conductivity vs. the temperature is observed for higher temperatures. Maximum conductivity values at 400 °C, 500 °C and 600 °C plotted as a function of the maximum exposure temperature are presented in Figure 4.

Figure 4 shows that layers containing Cu have roughly 4 times higher conductivity than the samples without. Also, an interesting maximum electrical conductivity for samples treated at 700 °C can be observed. It is only in the case of the MCO/Cu sample that a different trend is observed, possibly due to cracking. The highest overall electrical conductivity is found for the MCO/Cu sample. However, in this case, the conductivity values become irregular for temperatures higher than 600 °C, most likely due to the evolution of cracks in the film. The layer containing both the Cu and Fe also shows high electrical conductivity, whereas addition of Fe alone decreases the electrical conductivity.

From the observed temperature dependences of the electrical conductivity (Figure 3), activation energies were deduced from the expression [25]:

$$\sigma T = \sigma_0 \exp\left(-\frac{E_A}{k_B T}\right) \tag{3}$$

where: σ—electrical conductivity (S cm^{-1}), T—temperature (k), E—activation energy (eV), k_B—Boltzmann constant (eV k^{-1}).

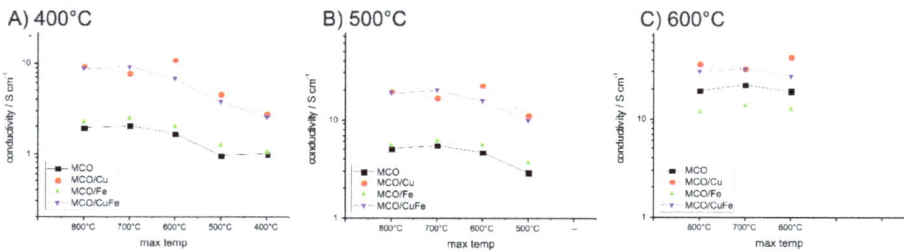

Figure 4. Electrical conductivity of layers as a function of maximum exposure temperature: (**A**) 400 °C, (**B**) 500 °C and (**C**) 600 °C.

Due to the inflection point on the conductivity curves, activation energies were calculated for a high (~800 °C–400 °C) and low (~400 °C–200 °C) temperature range (HT and LT). Calculated activation energies are presented in Table 2. For layers with the maximum processing temperature of 400 °C, the activation energy remains the same (~0.45 eV) for all materials. After the first heating to 500 °C, which is 100 °C above the deposition temperature, the activation energy of the low temperature regime lowers to ~0.40 eV for all samples. A visible variation is observed for the high temperature regime. The highest activation energy is reported for the MCO layer (~0.59 eV for the max temperature of 700 °C), with slightly lower values for the MCO/Fe layer (~0.52 eV for the max temperature of 700 °C). Both samples containing Cu have visibly lower activation energy, with the MCO/Cu having the lowest (~0.44 eV for a maximum temperature of 700 °C) and the MCO/CuFe being intermediate (~0.48 eV for the max temperature of 700 °C). The same value of the low temperature activation energy indicates a similar conduction mechanism. For high temperatures (400 °C–800 °C), substitution of Co with Cu clearly lowers the activation energy.

Table 2. Activation energy and electrical conductivity at 800°C for the coatings.

Name/S cm^{-1}/eV	σ 800 °C	E_A (max 800 °C)		E_A (max 700 °C)		E_A (max 600 °C)		E_A (max 500 °C)		E_A (max 400 °C)
		HT	LT	HT	LT	HT	LT	HT	LT	
MCO	73.0	0.58	0.40	0.59	0.40	0.57	0.41	0.54	0.42	0.45
MCO/Cu	85.2	0.43	0.46	0.44	0.41	0.40	0.38	0.44	0.40	0.44
MCO/Fe	37.9	0.52	0.41	0.52	0.40	0.54	0.41	0.54	0.43	0.44
MCO/CuFe	86.6	0.48	0.38	0.48	0.38	0.47	0.40	0.48	0.41	0.44

Electrical properties of the substituted (Mn,Co)$_3$O$_4$ spinels have been studied by other groups on bulk samples. Brylewski et al. [24] studied Cu$_x$Mn$_{1.25-0.5x}$Co$_{1.75-0.5x}$O$_4$ in the composition range $x = 0$ to 0.5. A change in the activation energy was also reported in this work. For the undoped spinel, activation energies of 0.68 eV and 0.41 eV were reported for the high and low temperature regions, respectively. The addition of Cu up to $x = 0.3$ lowered the activation energy in both the high temperature (0.58 eV) and low temperature regions (0.28 eV). The conductivity of the spinel at 800 °C equaled 162 S cm^{-1}.

Masi et al. have studied single and double substituted spinels (reported for the first time to the best knowledge of the authors) with compositions similar to the ones investigated in this work [13,26]. The high temperature activation energies were lowered by addition of the Cu. For $MnCo_{1.6}Fe_{0.4}O_4$, $MnCo_{1.8}Cu_{0.2}O_4$, $MnCo_{1.6}Cu_{0.2}Fe_{0.2}O_4$ the activation energies of 0.53 eV, 0.46 eV and 0.50 eV were reported, respectively. The electrical conductivity of $MnCo_{1.6}Cu_{0.2}Fe_{0.2}O_4$ has been reported to be ~82 S cm^{-1}, while for the undoped spinel ~75 S cm^{-1} and clearly the values are similar to the values reported here. Similarly, the addition of Fe leads to a decreased electrical conductivity. In summary, measurements reported here for thin films agree well with literature studies carried out on bulk samples.

3.3. Microstructural Characterization

To characterize the microstructure of the layers, X-ray diffractometry and scanning electron microscopy methods were applied.

All four layers were investigated by in-situ high temperature XRD. As produced layers were evaluated first during heating to 400 °C, 600 °C and 800 °C in order to determine their crystallite size and growth using Equation (1). Additionally, after heating to 800 °C and holding for 1 h, XRD patterns were recorded every 100 °C during cooling to check the change in lattice parameters according to the Equation (2). Exemplary results obtained for the $MnCo_{1.6}Cu_{0.4}O_4$ sample are shown in Figure 5 and summarized in Figure 6.

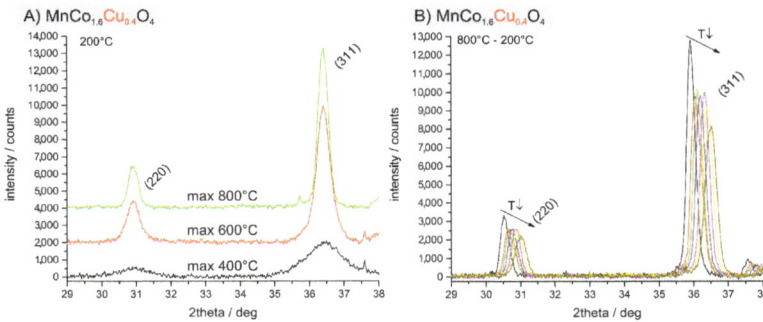

Figure 5. High temperature XRD of the MCO/Cu layer: spectra at 200 °C of layers after different max exposure temperature (**A**) and cooling cycle from 800 °C to 200 °C (**B**).

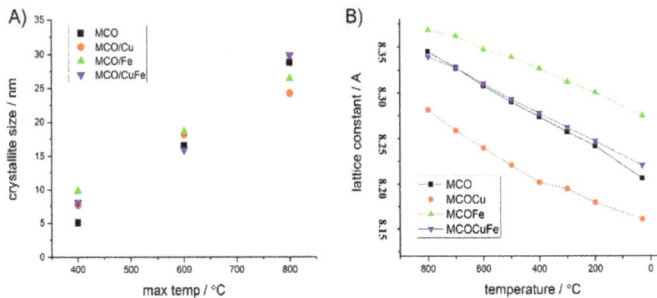

Figure 6. Crystallite size of the spinel (**A**) and lattice change due to temperature (**B**).

With an increase in maximum processing temperature, peaks become narrower and more intensive, as seen in Figures 5A and 6A; this is related to growing crystallite size, as presented in Figures 5A and 6A. For all samples, for the initial lowest processing temperature of 400 °C, peaks (220 and 311) are slightly shifted towards a higher 2θ range in comparison to layers processed at higher temperatures. This means that after the deposition at 390 °C, for very small crystallites (<10 nm), the lattice parameter is slightly smaller than after high temperature exposure. For layers processed at 600 °C and 800 °C, the lattice parameter stays the same, and only the crystallite size changes.

Crystallite growth evaluated by XRD peak broadening is similar for all layers. Crystallites grow from the initial ~6 nm to ~25 nm after processing at 800 °C. The Initial crystallite size is similar to the one observed previously by SEM in a similar system where $MnCo_2O_4$ precursors were impregnated into an $MnCo_2O_4$ matrix [11].

The variation in the lattice parameter (analyzing the structure as cubic) with temperature, measured during cooling from 800 °C is presented in Figure 6B. A clear effect of the introduced dopants on the lattice constant is observed. The addition of Cu lowers the lattice constant, whereas the addition of Fe increases the lattice constant. Simultaneous addition of the two dopants seems to cancel out the change in the lattice parameter. Observed changes are consistent with the ones reported in the literature [13,24,27].

XRD patterns measured at room temperature are presented in Figure 7. The different position of the 220 and 311 peaks represent a different cell size, as summarized in Figure 6B. Due to a low thickness of the film (~1 μm), the strongest peak comes from the sapphire substrate. After the heating to 800 °C, still only a single spinel phase is detected in the film.

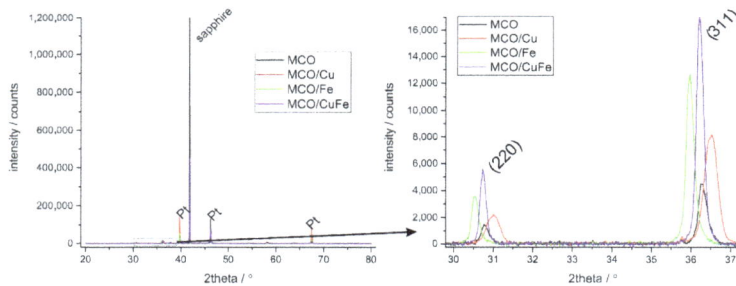

Figure 7. XRD at RT after exposure to 800 °C.

In an $MnCo_2O_4$ spinel, the tetrahedral sites are occupied preferentially by Co^{2+} cations, with octahedral sites occupied by mixed valence Co^{2+}, Co^{3+}, Mn^{3+}, Mn^{4+}. As summarized by Masi et al. [13], the addition of Fe in place of Co results in substitution in the octahedral position of Co^{2+} by Fe^{3+}, thus reducing the electrical conductivity. On the other hand, added Cu tends to occupy preferentially tetrahedral sites with the presence of Cu^+/Cu^{2+} which additionally promotes oxidation of Mn^{3+} to Mn^{4+} to maintain charge neutrality. This, in turn, increases the number of different valence species, possibly leading to higher electrical conductivity and thermal expansion. In general, the spinel structure is quite complex, as cations (Mn, Co, Fe, Cu) can have different oxidation states and additionally some spinel inversion is possible, where "B" atoms can occupy "A" sites and lead to mixed composition of the cation sublattices. Therefore, no single defect model is proposed here.

High magnification surface SEM images are presented in Figure 8. The samples are shown after the high temperature XRD evaluation with a maximum temperature of 800 °C for two hours. Well sintered and dense films are observed for all samples. There are evident differences in grain sizes. The smallest grains are found for the undoped spinel (49 ± 12 nm), followed by MCO/Fe (67 ± 14 nm) and with the largest grains for the Cu containing layers: MCO/CuFe with 126 ± 12 nm and MCO/Cu with 178 ± 10 nm grains. Both dopants increase the average grain size, with Cu having a stronger effect. Enhanced sintering of the Cu containing spinels has been one of the reasons (in addition to enhanced electrical conductivity) for the focus on them [28,29]. Grain sizes determined by SEM are different from the grain sizes determined by the XRD. One possible explanation might be that due to a large strain from the deposition method and from the substrate TEC mismatch, the Scherrer formula becomes inaccurate due to peak modification in the XRD spectra by strain in the simple procedure used. Very good surface coverage of the sapphire substrates has been achieved by the spray pyrolysis method with dense coatings prepared at a maximum temperature of only 800 °C. Temperatures required to achieve dense layers prepared by spray pyrolysis are typically 200 °C–400 °C lower than temperatures in standard powder processing methods.

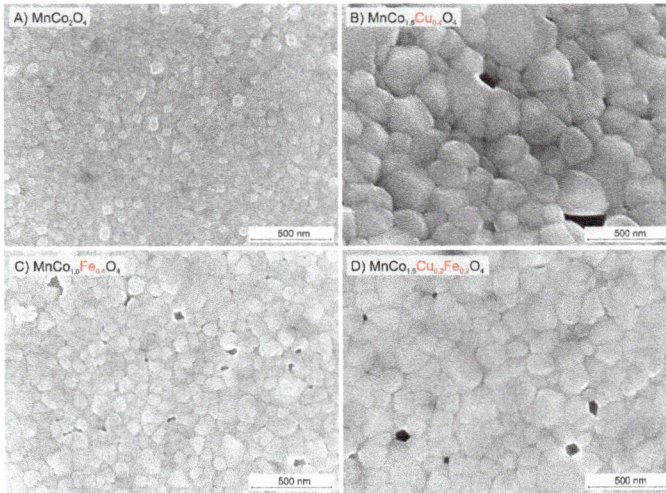

Figure 8. (**A–D**) SEM surface view of all layers after the electrical conductivity test.

A tilted (37°) high-magnification image of a fracture cross section of the $MnCo_{1.6}Cu_{0.4}O_4$ layer is shown in Figure 9A alongside lower magnification pictures of the fractured films. Thickness varies between ~0.7 μm for the $MnCo_2O_4$ to 1.2 μm for the $MnCo_{1.6}Cu_{0.2}Fe_{0.2}O_4$. The thickness of each coating is uniform over the whole cross section.

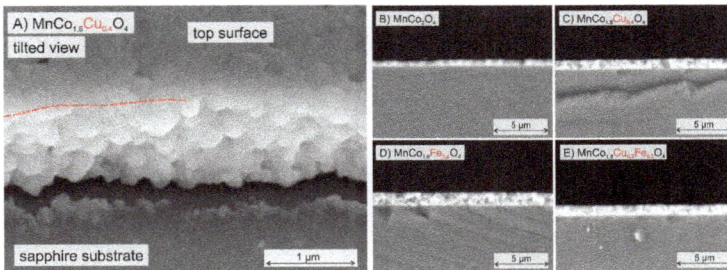

Figure 9. (**A–E**) Fracture cross section SEM image of the MCO/Cu layer after the electrical conductivity test. Picture (**A**) taken at an angle of 37°.

The thickness of the $MnCo_{1.6}Cu_{0.4}O_4$ layer is built of approximately 10 grains. No large pores are observed in the cross section. Based on the surface and cross-section images, it seems that spray pyrolysis is a reliable and effective method for deposition of 1 μm thick layers of different compositions. In the previous deposition studies, it was not possible to obtain layers thicker than 0.3–0.5 μm in a single step process [30,31]. Increasing the molarity of the precursor solution allowed for obtaining thicker layers for the same deposition time, extending possible applications for the spray pyrolysis deposition method.

SEM pictures of the surface of the MCO/Cu layer are presented in Figure 10. Cracks are clearly visible over most of the sample surface. Both of the two MCO/Cu samples evaluated were cracked. In some places the layer spalled off; no cracks were observed for the other layers. These cracks were not detected in the as-produced sample; they must have formed during heat treatment of the layers. As previously described, irregularities in electrical conductivity measurements for the MCO/Cu layers were found. It is likely that the cracking of the layer is responsible for the step-changes in the measured electrical conductivity seen in Figures 3 and 4.

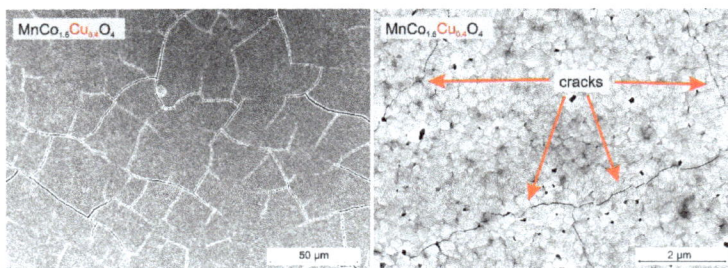

Figure 10. SEM surface images showing cracks on the MCO/Cu layer after the electrical conductivity test.

Substitution of Co by Cu leads to an increase in the thermal expansion coefficient. According to Masi et al. [13] and Brylewski et al. [24], the thermal expansion coefficient (TEC between 30 °C–800 °C) of a partially Cu substituted spinel can reach ~15 ppm K^{-1}. For partial Fe substitution a lowering of TEC is observed (to ~12.5 ppm K^{-1}) and for the undoped spinel values between 12.5 and 14 are typically reported [13]. Sapphire substrate has a much lower TEC (~7.5 ppm K^{-1}), so the stresses caused by the TEC mismatch are quite high. However, MCO/Cu and MCO/CuFe should have quite similar TEC. Other factors possibly influence the risk of cracking. The risk of crack formation further increases with film thickness, which varies somewhat between the samples. Cracking might also be influenced by grain growth and mass transport. Enhanced sintering of the Cu doped sample will cause lateral stress on the constrained layer and increase the risk of cracking. Simultaneous addition of Cu and Fe does not lead to such rapid grain growth and therefore doubly doped composition offers an important stability advantage.

4. Summary and Conclusions

High quality ~1 μm thick layers of $MnCo_2O_4$, $MnCo_{1.6}Cu_{0.4}O$, $MnCo_{1.6}Fe_{0.4}O_4$ and $MnCo_{1.6}Cu_{0.2}Fe_{0.2}O_4$ have been prepared by spray pyrolysis at 390 °C. Electrical conductivity and microstructural changes of layers have been evaluated up to 800 °C. Spray pyrolysis is proven to produce dense spinel layers at temperatures of only 800 °C, which is lower than for standard ceramic processing methods. Maximum electronic conductivity is found for layers experiencing a maximum processing temperature of 700 °C, whereas no changes in the activation energy were noticed for different maximum processing temperatures. Electrical conductivity values obtained for the films agree well with the values reported for bulk samples. The addition of Cu increases the electrical conductivity and decreases the activation energy at high temperatures whereas the addition of Fe decreases the electrical conductivity. The addition of Cu increases the lattice parameter whereas the addition of Fe lowers the lattice parameter and for a Fe, Cu co-substituted spinel, lattice changes cancel out and resemble the undoped structure. Additionally, Cu visibly enhances grain growth and leads to cracked films. Simultaneous addition of Cu and Fe seems advantageous and offer the benefits of higher electrical conductivity and limited grain growth.

Acknowledgments: DTU Energy acknowledges support from the project 2015-1-12276 "Towards solid oxide electrolysis plants in 2020", ForskEL, energienet.dk, while GUT acknowledges partly support from the project DZP/PL-TW2/6/2015 "Innovative Solid Oxide Electrolyzers for Storage of Renewable Energy". Authors acknowledges for Statutory Funds for Research of GUT.

Author Contributions: Dagmara Szymczewska prepared the layers by spray pyrolysis and measured their electrical conductivity and wrote parts of the manuscript. S.M. analyzed the samples by SEM and XRD and wrote most of the manuscript. Peter Vang Hendriksen corrected the manuscript and discussed the research plan and the results. Piotr Jasinski corrected the manuscript and lead the discussions of the results.

Conflicts of Interest: The authors declare no conflict of interest.

References

1. Ginley, T.; Wang, Y.; Law, S. Topological insulator film growth by molecular beam epitaxy: A review. *Crystals* **2016**, *6*, 154. [CrossRef]
2. Deng, L.; Wang, K.; Zhao, C.X.; Yan, H.; Britten, J.F.; Xu, G. Phase and texture of solution-processed copper phthalocyanine thin films investigated by two-dimensional grazing incidence X-ray diffraction. *Crystals* **2011**, *1*, 112–119. [CrossRef]
3. Lin, D.; Li, Z.; Li, F.; Cai, C.; Liu, W.; Zhang, S. Tetragonal-to-tetragonal phase transition in lead-free ($k_x na_{1-x}$) nbo3 ($x = 0.11$ and 0.17) crystals. *Crystals* **2014**, *4*, 113–122. [CrossRef]
4. Chen, Y.; Santos, D.M.F.; Sequeira, C.A.C.; Lobo, R.F.M. Studies of modified hydrogen storage intermetallic compounds used as fuel cell anodes. *Crystals* **2012**, *2*, 22–33. [CrossRef]
5. Molenda, J.; Kupecki, J.; Baron, R.; Blesznowski, M.; Brus, G.; Brylewski, T.; Bucko, M.; Chmielowiec, J.; Cwieka, K.; Gazda, M.; et al. Status report on high temperature fuel cells in Poland—Recent advances and achievements. *Int. J. Hydrogen Energy* **2017**, *42*, 4366–4403. [CrossRef]
6. Wu, X.; Li, S.; Wang, B.; Liu, J.; Yu, M. Controllable synthesis of micro/nano-structured $MnCo_2O_4$ with multiporous core–shell architectures as high-performance anode materials for lithium-ion batteries. *New J. Chem.* **2015**, *39*, 8416–8423. [CrossRef]
7. Jin, Y.; Wang, L.; Jiang, Q.; Du, X.; Ji, C.; He, X. Mesoporous $MnCo_2O_4$ microflower constructed by sheets for lithium ion batteries. *Mater. Lett.* **2016**, *177*, 85–88. [CrossRef]
8. Akhtar, M.A.; Sharma, V.; Biswas, S.; Chandra, A. Tuning porous nanostructures of $MnCo_2O_4$ for application in supercapacitors and catalysis. *RSC. Adv.* **2016**, *6*, 96296–96305. [CrossRef]
9. Larring, Y.; Norby, T. Spinel and perovskite functional layers between plansee metallic interconnect (Cr-5 wt % Fe-1 wt % Y_2O_3) and ceramic ($La_{0.85}Sr_{0.15})_{0.91}MnO_3$ cathode materials for solid oxide fuel cells. *J. Electrochem. Soc.* **2000**, *147*, 3251–3256. [CrossRef]
10. Talic, B.; Falk-Windisch, H.; Venkatachalam, V.; Hendriksen, P.V.; Wiik, K.; Lein, H.L. Effect of coating density on oxidation resistance and Cr vaporization from solid oxide fuel cell interconnects. *J. Power Sources* **2017**, *354*, 57–67. [CrossRef]
11. Molin, S.; Jasinski, P.; Mikkelsen, L.; Zhang, W.; Chen, M.; Hendriksen, P.V. Low temperature processed $MnCo_2O_4$ and $MnCo_{1.8}Fe_{0.2}O_4$ as effective protective coatings for solid oxide fuel cell interconnects at 750 °C. *J. Power Sources* **2016**, *336*, 408–418. [CrossRef]
12. Montero, X.; Tietz, F.; Sebold, D.; Buchkremer, H.P.; Ringuede, A.; Cassir, M.; Laresgoiti, A.; Villarreal, I. $MnCo_{1.9}Fe_{0.1}O_4$ spinel protection layer on commercial ferritic steels for interconnect applications in solid oxide fuel cells. *J. Power Sources* **2008**, *184*, 172–179. [CrossRef]
13. Masi, A.; Bellusci, M.; McPhail, S.J.; Padella, F.; Reale, P.; Hon, J.E.; Wilckens, R.S.; Carlini, M. The effect of chemical composition on high temperature behaviour of Fe and Cu doped Mn-Co spinels. *Ceram. Int.* **2017**, *43*, 2829–2835. [CrossRef]

14. Shaigan, N.; Qu, W.; Ivey, D.G.; Chen, W. A review of recent progress in coatings, surface modifications and alloy developments for solid oxide fuel cell ferritic stainless steel interconnects. *J. Power Sources* **2010**, *195*, 1529–1542. [CrossRef]

15. Wang, K.; Liu, Y.; Fergus, J.W. Interactions between SOFC interconnect coating materials and chromia. *J. Am. Ceram. Soc.* **2011**, *94*, 4490–4495. [CrossRef]

16. Jasinski, P. Electrical properties of nanocrystalline Sm-doped ceria ceramics. *Solid State Ion.* **2006**, *177*, 2509–2512. [CrossRef]

17. Rupp, J.L.M.; Gauckler, L.J. Microstructures and electrical conductivity of nanocrystalline ceria-based thin films. *Solid State Ion.* **2006**, *177*, 2513–2518. [CrossRef]

18. Rupp, J.L.M.; Infortuna, A.; Gauckler, L.J. Microstrain and self-limited grain growth in nanocrystalline ceria ceramics. *Acta Mater.* **2006**, *54*, 1721–1730. [CrossRef]

19. Molin, S.; Jasinski, P.Z. Improved performance of LaNi$_{0.6}$Fe$_{0.4}$O$_3$ solid oxide fuel cell cathode by application of a thin interface cathode functional layer. *Mater. Lett.* **2017**, *189*, 252–255. [CrossRef]

20. Molin, S.; Chrzan, A.; Karczewski, J.; Szymczewska, D.; Jasinski, P. The role of thin functional layers in solid oxide fuel cells. *Electrochim. Acta* **2016**, *204*, 136–145. [CrossRef]

21. Jasinski, P.; Molin, S.; Gazda, M.; Petrovsky, V.; Anderson, H.U. Applications of spin coating of polymer precursor and slurry suspensions for Solid Oxide Fuel Cell fabrication. *J. Power Sources* **2009**, *194*, 10–15. [CrossRef]

22. Szymczewska, D.; Molin, S.; Chen, M.; Hendriksen, P.V.; Jasinski, P. Ceria based protective coatings for steel interconnects prepared by spray pyrolysis. *Proc. Eng.* **2014**, *98*, 93–100. [CrossRef]

23. Brylewski, T.; Kucza, W.; Adamczyk, A.; Kruk, A.; Stygar, M.; Bobruk, M.; Dąbrowa, J. Microstructure and electrical properties of Mn$_{1+x}$Co$_{2-x}$O$_4$ ($0 \leq x \leq 1.5$) spinels synthesized using EDTA-gel processes. *Ceram. Int.* **2014**, *40*, 13873–13882. [CrossRef]

24. Brylewski, T.; Kruk, A.; Bobruk, M.; Adamczyk, A.; Partyka, J.; Rutkowski, P. Structure and electrical properties of Cu-doped Mn-Co-O spinel prepared via soft chemistry and its application in intermediate-temperature solid oxide fuel cell interconnects. *J. Power Sources* **2016**, *333*, 145–155. [CrossRef]

25. Maier, J. *Physical Chemistry of Ionic Materials: Ions and Electrons in Solids*; Wiley: Stuttgart, Germany, 2004.

26. Masi, A.; Bellusci, M.; McPhail, S.J.; Padella, F.; Reale, P.; Hong, J.; Robert, S.-W.; Carlini, M. Cu-Mn-Co oxides as protective materials in SOFC technology: The effect of chemical composition on mechanochemical synthesis, sintering behaviour, thermal expansion and electrical conductivity. *J. Eur. Ceram. Soc.* **2017**, *37*, 661–669. [CrossRef]

27. Talic, B. Metallic Interconnects for Solid Oxide Fuel Cells: High Temperature Corrosion and Protective Spinel Coatings. Ph.D. Thesis, Norwegian University of Science and Technology, Trondheim, Norway, 2016. Available online: http://hdl.handle.net/11250/2404554 (accessed on 19 May 2017).

28. Chen, G.; Xin, X.; Luo, T.; Liu, L.; Zhou, Y.; Yuan, C.; Lin, C.; Zhan, Z.; Wang, S. Mn$_{1.4}$Co$_{1.4}$Cu$_{0.2}$O$_4$ spinel protective coating on ferritic stainless steels for solid oxide fuel cell interconnect applications. *J. Power Sources* **2015**, *278*, 230–234. [CrossRef]

29. Bobruk, M.; Durczak, K.; Dąbek, J.; Brylewski, T. Structure and electrical properties of Mn-Cu-O Spinels. *J. Mater. Eng. Perform.* **2017**, *26*, 1598–1604. [CrossRef]

30. Szymczewska, D.; Chrzan, A.; Karczewski, J.; Molin, S.; Jasinski, P. Spray pyrolysis of doped-ceria barrier layers for solid oxide fuel cells. *Surf. Coat. Technol.* **2017**, *313*, 168–176. [CrossRef]

31. Szymczewska, D.; Karczewski, J.; Chrzan, A.; Jasinski, P. CGO as a barrier layer between LSCF electrodes and YSZ electrolyte fabricated by spray pyrolysis for solid oxide fuel cells. *Solid State Ion.* **2017**, *302*, 113–117. [CrossRef]

© 2017 by the authors. Licensee MDPI, Basel, Switzerland. This article is an open access article distributed under the terms and conditions of the Creative Commons Attribution (CC BY) license (http://creativecommons.org/licenses/by/4.0/).

crystals

MDPI

Article

Element Strategy Using Ru-Mn Substitution in CuO-CaCu$_3$Ru$_4$O$_{12}$ Composite Ceramics with High Electrical Conductivity

Akihiro Tsuruta [1,*], Masashi Mikami [1], Yoshiaki Kinemuchi [1], Ichiro Terasaki [1,2], Norimitsu Murayama [3] and Woosuck Shin [1]

[1] National Institute of Advanced Industrial Science and Technology (AIST), Shimo-Shidami, Moriyama-ku, Nagoya 463-8560, Japan; m-mikami@aist.go.jp (M.M.); y.kinemuchi@aist.go.jp (Y.K.); terra@cc.nagoya-u.ac.jp (I.T.); w.shin@aist.go.jp (W.S.)

[2] Department of Physics, Nagoya University, Furo-cho, Chuikusa-ku, Nagoya 464-8602, Japan

[3] National Institute of Advanced Industrial Science and Technology (AIST), 1-1-1 Higashi, Tsukuba 305-8565, Japan; n-murayama@aist.go.jp

* Correspondence: a.tsuruta@aist.go.jp; Tel.: +81-52-736-7481

Academic Editor: Stevin Snellius Pramana
Received: 15 May 2017; Accepted: 8 July 2017; Published: 10 July 2017

Abstract: CaCu$_3$Ru$_{4-x}$Mn$_x$O$_{12}$ bulks with various substitution amounts x and sintering additive CuO (20 vol.%) were prepared, and the influence of x on the electrical conductivity in a wide temperature range (8–900 K) was investigated. Microstructural observations showed an enhancement of bulk densification upon Mn substitution. Although the resistivity increased with increasing x, the resistivity was as low as a few mΩcm even in the sample with $x = 2.00$, where half of Ru is substituted by Mn. This high conductivity despite the loss of Ru 4d conduction following the substitution is explained by the A-site (Cu^{2+}) conduction in CaCu$_3$Ru$_{4-x}$Mn$_x$O$_{12}$. The thermopower of CaCu$_3$Ru$_{4-x}$Mn$_x$O$_{12}$ was found to be influenced by the substitution, and a sign inversion was observed in the substituted samples at low temperature. The partial substitution of Ru by Mn in CaCu$_3$Ru$_4$O$_{12}$ enables the reduction of the materials cost while maintaining good electrical conductivity for applications as a conducting device component.

Keywords: conducting oxide; composite; perovskite; substitution

1. Introduction

Perovskites of the general formula AC$_3$B$_4$O$_{12}$ represent a large family of materials and can be considered as a fourfold superstructure of the ABO$_3$ perovskite in which a cation (A-site) and Jahn-Teller ions (C-site; Cu^{2+}, Mn^{3+}) are long-range ordered in a double-cubic unit cell. A wide variety of cations can occupy the A-, B- and C-sites, and various partial substitutions are possible on each of these sites [1–6]. Many interesting properties have been found in these compounds, leading to potential applications in various fields. For instance, CaCu$_3$Ti$_4$O$_{12}$ and CaMn$_{3-x}$Cu$_x$Mn$_4$O$_{12}$ show an anomalously high dielectric constant and a giant magneto-resistance, respectively [4,5,7–9]; LaCu$_3$Fe$_4$O$_{12}$ is a negative thermal expansion material [10]; and CaCu$_3$Ru$_4$O$_{12}$ exhibits high metallic electrical conductivity [11,12]. Figure 1 shows the crystal structure of CaCu$_3$Ru$_4$O$_{12}$. Ca^{2+} and Cu^{2+} share the A-site of the perovskite. The Ru-ions (Ru^{4+}) occupy the B-site at the point (1/4, 1/4, 1/4) of the lattice. The O-ions (O^{2-}) are shifted from their regular position at (1/4, 1/4, 0) in simple perovskites as the RuO$_6$ octahedra are tilted [13].

Recently, many high-temperature operating electrical devices such as gas sensors and solid oxide fuel cells have been actively developed. Their conductive materials are required to be stable at high temperatures and in various atmospheres [14–20]. Platinum is conventionally used as the conductive materials in such devices, but its high cost has been a serious barrier to their widespread use. Owing to their stability at high temperatures, oxides appear as a potential alternative to platinum, provided they exhibit low resistivity. Among the conducting metal oxides, noble-metal oxides such as RuO_2 [21], IrO_2 [21,22], and ReO_2 [23] exhibit the lowest resistivity. Next comes $CaCu_3Ru_4O_{12}$, with a resistivity value at room temperature below 0.5 mΩcm, which is one order of magnitude lower than that of La-Sr-Co-Fe-O perovskites (over 10 mΩcm) at room temperature [24]. Therefore, $CaCu_3Ru_4O_{12}$ is a suitable candidate to replace platinum in devices. However, there are two main challenges with the use of $CaCu_3Ru_4O_{12}$ in devices. One is its resistance to sintering, but we have found that a possible solution to this issue is to use CuO as a sintering additive [25]. The other is the high cost of Ru. Fortunately, various alternative elements are expected to substitute Ru on the B-site to reduce the material cost of $CaCu_3Ru_4O_{12}$ while keeping its resistivity low. In addition, the electrical conducting mechanism in $CaCu_3Ru_4O_{12}$ has been investigated by Kobayashi et al [11], who have shown that not only the Ru-O network, but also Cu^{2+} through Kondo coupling [9] between Ru 4d and Cu 3d electrons contributes to the good electrical conduction of the material. This result encourages our scheme to reduce the amount of Ru in $CaCu_3Ru_4O_{12}$ by substitution while maintaining a high electrical conductivity. In order to use conducting oxides as substitute materials for platinum, their resistivity should be a few mΩcm over 500 °C and show a temperature dependence similar to that of metal.

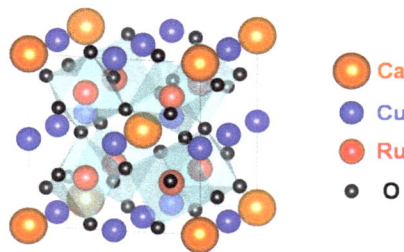

Figure 1. Crystal structure of $CaCu_3Ru_4O_{12}$. (The figure was drawn using VESTA [26].)

In this study, we focused on Mn as a substitute element for Ru in $CaCu_3Ru_4O_{12}$ with a fixed amount of sintering additive of 20 vol.% CuO. While the crystal structure and magnetism of $CaCu_3Ru_{4-x}Mn_xO_{12}$ have been discussed in a previous report [6], no detailed information on the conducting properties of the material is available. Here, the transport properties of 20 vol.% CuO-mixed $CaCu_3Ru_{4-x}Mn_xO_{12}$ with various Mn substitution amounts (in the range 0.00–2.00) were investigated over a wide temperature range (8–900 K).

2. Experimental Procedure

$CaCu_3Ru_{4-x}Mn_xO_{12}$ powders with various Mn substitution amounts (x = 0.00, 0.25, 0.50, 0.75, 1.00, 1.25, 1.50, 1.75 and 2.00) were prepared by a solid-state reaction method. Stoichiometric mixtures of $CaCO_3$, CuO, RuO_2, and Mn_3O_4 were pressed into pellets and calcined in air at 1000 °C for 48 h. During calcination, the pellets were surrounded by a mixture of $CaCO_3$, CuO, and RuO_2 powders to prevent the sublimation of Ru and a consequent composition deviation. In our preliminary experiment, Ru sublimation and formation of second phases such as $CaRuO_3$ were confirmed after calcination at 1000 °C without the surroundings. In addition, although the exact solubility limit of Mn at the Ru-site obtained with a synthesis at atmospheric pressure is unknown, it was not possible to synthesize $CaCu_3RuMn_3O_{12}$ (x = 3) with this synthesis method.

$CaCu_3Ru_{4-x}Mn_xO_{12}$ powders were obtained via the mechanical grinding of the calcined pellets. The $CaCu_3Ru_{4-x}Mn_xO_{12}$ powders were then mixed with CuO powder acting as a sintering additive, pressed into a pellet, and sintered at 1000 °C for 48 h in air. The obtained samples are hereafter referred to as CuO(20 vol.%)-CaCu$_3$Ru$_{4-x}$Mn$_x$O$_{12}$ bulks. A CuO volume fraction of 20 vol.% was calculated using the molecular weights and lattice constants for each Mn substitution amount. Twenty vol.% of the samples with x = 0.00, 0.25, 0.50, 0.75, 1.00, 1.25, 1.50, 1.75, and 2.00 correspond to 19.6, 19.7, 19.8, 19.9, 20.0, 20.1, 20.1, 20.2, and 20.4 wt%, respectively.

X-ray diffraction (XRD) of the bulk samples was carried out using a standard diffractometer with CuKα radiation, in the 2θ-θ scan mode (Rigaku SmartLab, Tokyo, Japan). The morphology of the bulk samples was observed using a field emission scanning electron microscope (FE-SEM; JEOL JSM-6335FM, Tokyo, Japan). The resistivity and thermopower were measured from 7 K to 350 K using a conventional four-probe method and a steady-state technique in vacuum using a cryostat, and from 350 K to 900 K by a four-point probe method in air using an electrical conductivity and Seebeck coefficient measurement system (Ozawa Science RZ2001S, Nagoya, Japan). Since we used different methods and devices for the resistivity measurements below and above 350 K, the value of the resistivity at 350 K obtained with each method differed slightly for each sample. The error is considered to be due to the measurement error of the distance between the voltage terminals in the low-temperature measurements, and the difference between the high-temperature and low-temperature measurement values was less than 10%. Consequently, we normalized the low-temperature data so that the value found at 350 K by the low-temperature method matches that found at 350 K by the high-temperature method. An excellent agreement between the derivative of the two datasets at 350 K was found in all samples.

3. Results and Discussion

Figure 2a shows the XRD patterns of CuO(20 vol.%)-CaCu$_3$Ru$_{4-x}$Mn$_x$O$_{12}$ bulks for x varying between 0.00 and 2.00. All the patterns are similar, displaying peaks that can be assigned to CaCu$_3$Ru$_4$O$_{12}$ phase and CuO, respectively. The peaks corresponding to diffraction by the (440) planes of CaCu$_3$Ru$_4$O$_{12}$ phase are enlarged in Figure 2b. The peaks are systematically shifted to higher diffraction angles with increasing x, indicating that the lattice constant of the sample decreased as x increased. The lattice constant, which was calculated from the 2θ angles corresponding to (n00) planes, is plotted as a function of x in Figure 3. The lattice constant systematically decreases from 7.428 Å for CaCu$_3$Ru$_4$O$_{12}$ to 7.393 Å for CaCu$_3$Ru$_2$Mn$_2$O$_{12}$ with increasing x. As a charge transfer from (Ru^{4+} + Mn^{4+}) to (Ru^{5+} + Mn^{3+}) has been reported to take place in SrMn$_{1-y}$Ru$_y$O$_3$ [27], we should consider this possibility in CaCu$_3$Ru$_{4-x}$Mn$_x$O$_{12}$ as well. In both cases, given the respective ion size (Ru^{4+}: 0.62 Å, Ru^{5+}: 0.565 Å, Mn^{3+}: 0.645 Å, and Mn^{4+}: 0.53 Å), the shift observed in the XRD spectra can be explained by the smaller average size of the ions on the B-site of the substituted material (either Mn^{4+} and Ru^{4+}, or Mn^{3+} and Ru^{5+}) with respect to that of Ru^{4+}, even if the size reduction effect would be larger without the charge transfer. By contrast, the influence of such a charge transfer on the electrical transport and resistivity would be dramatically large. In light of our resistivity measurements discussed below, which show no such dramatic change upon substitution, we can infer that the valence of Ru did not change.

Figure 4 shows the XRD patterns of as-synthesized CaCu$_3$Ru$_4$O$_{12}$ and CaCu$_3$Ru$_2$Mn$_2$O$_{12}$ powders. The patterns of both are almost the same, and there are no peaks corresponding to the raw materials or some second phases. We accurately weighed the raw materials according to the stoichiometry of CaCu$_3$Ru$_2$Mn$_2$O$_{12}$. Therefore, if we assume that Mn could substitute Cu, yielding CaCu$_{3-y}$Mn$_y$Ru$_{4-x}$Mn$_x$O$_{12}$, then the composition ratio of the raw materials would not match that of the formed compound, and some peaks corresponding to the raw materials and/or second phases would be observed in the XRD patterns. From this result, we consider that a very small quantity of Mn may have substituted Cu. In Figure 2b, the peaks corresponding to the (440) plane of CaCu$_3$Ru$_{2.25}$Mn$_{1.75}$O$_{12}$ and CaCu$_3$Ru$_2$Mn$_2$O$_{12}$ are broader than those of the other samples. This broadening may due to a phase separation between the phases with and without Cu-Mn substitution in the samples with high Mn substitution.

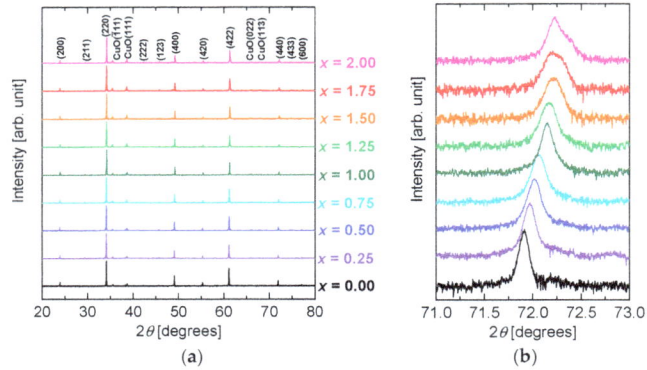

Figure 2. XRD (CuKα) patterns of the CuO(20 vol.%)-CaCu$_3$Ru$_{4-x}$Mn$_x$O$_{12}$ bulks. (**a**) Patterns between 20° and 80°. The peak labels without 'CuO' correspond to the CaCu$_3$Ru$_4$O$_{12}$ phase. (**b**) Enlarged patterns (71°–73°) for the peaks corresponding to diffraction by the (440) planes of the CaCu$_3$Ru$_4$O$_{12}$ phase.

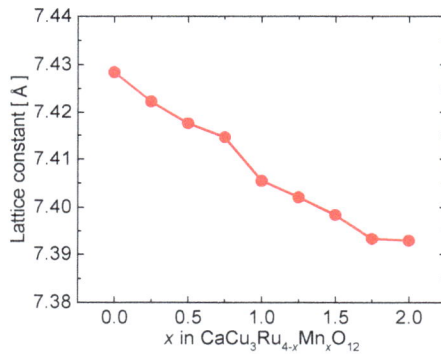

Figure 3. Lattice constant of CaCu$_3$Ru$_{4-x}$Mn$_x$O$_{12}$ in the CuO(20 vol.%)-CaCu$_3$Ru$_{4-x}$Mn$_x$O$_{12}$ bulks plotted as a function of Mn substitution amount x.

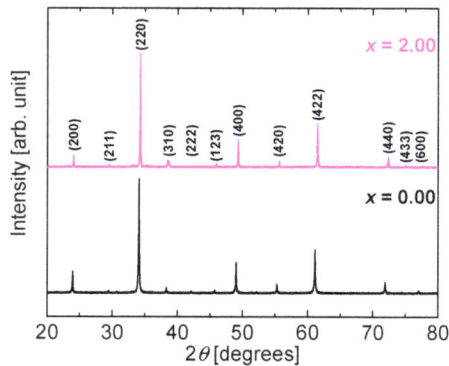

Figure 4. XRD (CuKα) patterns of the as-synthesized CaCu$_3$Ru$_4$O$_{12}$ and CaCu$_3$Ru$_2$Mn$_2$O$_{12}$ powders.

Figure 5a through c show FE-SEM images of the surface of the samples with $x = 0.00$, 0.75, and 1.50, respectively. In all samples, the grains fused together, and there are a lot of voids. The diameter of the grain was about 3 and 5 μm without and with Mn substitution, respectively. The grain size in the samples with Mn substitution was observed to be almost the same for all values of $x \neq 0.00$. As it depends on the grain size, the thickness of the necks between the grains was larger in the substituted samples than in the unsubstituted sample. Figure 5d shows the relative density of CuO(20 vol.%)-CaCu$_3$Ru$_{4-x}$Mn$_x$O$_{12}$ bulks as a function of x. The relative density displayed a dramatic increase between $x = 0.00$ (unsubstituted sample) and $x = 0.25$, and then slightly increased with increasing x. The results shown in Figure 5 suggest that Mn substitution increased the defect density, which promoted volume diffusion, and, in turn, grain growth and densification of the CaCu$_3$Ru$_{4-x}$Mn$_x$O$_{12}$ phase.

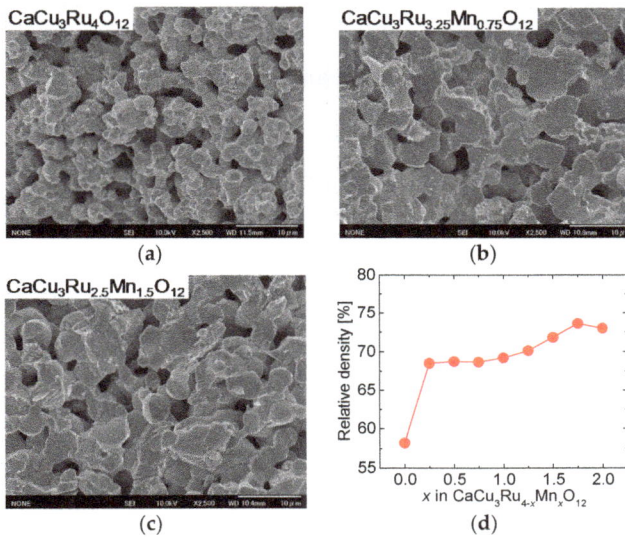

Figure 5. FE-SEM images of the surface of the CuO(20 vol.%)-CaCu$_3$Ru$_{4-x}$Mn$_x$O$_{12}$ bulks with $x = 0.00$ (**a**), 0.75 (**b**), and 1.50 (**c**). (**d**) Relative density of the samples as a function of x.

The temperature dependence of the resistivity (ρ-T curve) of the CuO(20 vol.%)-CaCu$_3$Ru$_{4-x}$Mn$_x$O$_{12}$ bulks is shown in Figure 6. The resistivity systematically increased with increasing x except for the sample with $x = 0.00$, which is attributed to its lower relative density with respect to the other (Mn-substituted) samples. A CaCu$_3$Ru$_4$O$_{12}$ bulk sample prepared with a similar relative density to that of the Mn-substituted samples would be expected to have a lower resistivity than that of the sample with $x = 0.25$. It is noteworthy that the resistivity value was always as low as a few mΩcm, even in the sample with $x = 2.00$, which contains as many Mn as Ru ions. In the case of other perovskite materials with Ru-Mn substitution such as CaRu$_{1-x}$Mn$_x$O$_3$ and SrRu$_{1-x}$Mn$_x$O$_3$, the substitution of half of the Ru-sites with Mn causes a drastic increase in the resistivity or band gap [28,29]. In particular, the resistivity increases at 300 K from 2 mΩcm for SrRuO$_3$ to 20 mΩcm for SrRu$_{0.5}$Mn$_{0.5}$O$_3$. In these materials, Ru 4d electrons are responsible for electrical conduction, while Mn substitution reduces the overlaps between the electron clouds. However, as mentioned in the introduction, in the case of CaCu$_3$Ru$_4$O$_{12}$, electrons from the A-site (Cu^{2+}) are also responsible for electrical conduction [11], which explains the low resistivity of CaCu$_3$Ru$_2$Mn$_2$O$_{12}$.

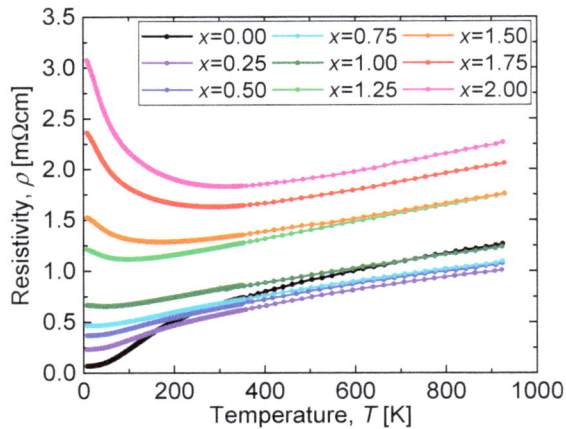

Figure 6. Temperature dependence of the resistivity of CuO(20 vol.%)-CaCu$_3$Ru$_{4-x}$Mn$_x$O$_{12}$ bulks.

While the resistivity systematically increased with increasing x, it showed the same temperature dependence at high temperatures for all values of x. The reason for this is that phonon scattering following Matthiessen's rule is the dominant parameter that influences the electrical resistivity at temperatures above 300 K. On the other hand, the temperature dependence of the resistivity at low temperatures differed according to the value of x. The residual resistivity $\rho(T = 0)$ increased with increasing x below $x = 1.00$. Above $x = 1.25$, the resistivity increased with cooling, which corresponds to a semiconducting behavior.

Let us analyze the semiconducting behavior at low temperatures of Mn-substituted samples using two kinds of transport mechanism, the activated conduction and the hopping conduction, which are representative conduction mechanisms for semiconducting electrical transport. Parts of the ρ-T curves corresponding to a transport mechanism dominated by activated conduction can be fitted by the following equation:

$$\rho = \rho_0 \exp(E_g/k_B T), \tag{1}$$

where k_B, T, and E_g are the Boltzmann constant, absolute temperature, and activation energy, respectively. In that way, the activation energy, i.e., the transport gap E_g, can be determined. Figure 7a shows $\ln(\rho/m\Omega cm)$ plotted against T^{-1} for CaCu$_3$Ru$_2$Mn$_2$O$_{12}$. The fitting curve corresponding to Equation (1) is shown as a broken line. The plot was well fitted by Equation (1) between $T^{-1} = 5.563 \times 10^{-3}$ (i.e., $T = 179.76$ K) and 4.204×10^{-3} (i.e., $T = 237.86$ K), in good agreement with the minimal value of the resistivity showed in Figure 6. From the fitting curve, the activation energy, expressed as a temperature, was determined as $E_g/k_B = 30.28$ K. This temperature is much lower than the temperature at which semiconducting behavior begins, indicating that activated conduction is not employed in CaCu$_3$Ru$_2$Mn$_2$O$_{12}$ at low temperatures; hopping conduction is the conduction mechanism at work.

Variable range hopping (VRH) conduction is one type of hopping conduction that gives rise to a temperature dependence of the resistivity of the form [30]:

$$\rho = \rho_0 \exp(U/T^{1/4}). \tag{2}$$

Figure 7b shows $\ln(\rho/m\Omega cm)$ plotted against $T^{-1/4}$ for CaCu$_3$Ru$_2$Mn$_2$O$_{12}$. The broken line is the fitting curve obtained with Equation (2). The fitting line fits to the plot for $T^{-1/4}$ between 0.3612 (i.e., $T = 58.76$ K) and 0.2609 (i.e., $T = 215.89$ K). This temperature range roughly corresponds to the temperature region where CaCu$_3$Ru$_2$Mn$_2$O$_{12}$ showed semiconducting behavior according to the ρ-T curve shown in Figure 6. Accordingly, it can be concluded that the conducting mechanism in $CaCu_3Ru_2Mn_2O_{12}$ at low temperatures is VRH conduction. Hopping conduction occurs due to the discontinuity of Ru conduction caused by Mn substitution, which explains the expansion of the temperature range of the semiconducting behavior to higher temperatures with increasing x, as shown in Figure 6.

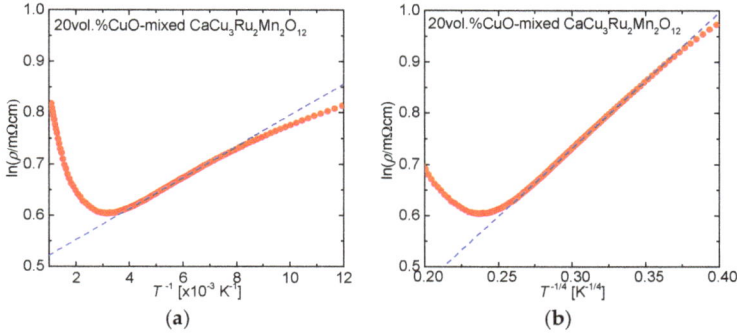

Figure 7. Temperature dependence of the resistivity of CuO(20 vol.%)-$CaCu_3Ru_2Mn_2O_{12}$ (**a**) Plot of $\ln(\rho/m\Omega cm)$ against T^{-1} (**b**) Plot of $\ln(\rho/m\Omega cm)$ against $T^{-1/4}$. The broken lines correspond to active conduction (**a**) and variable range hopping (**b**) as given by Equations (1) and (2), respectively.

From these results, it appears that two electrical conduction mechanisms underlie the electrical conduction in $CaCu_3Ru_{4-x}Mn_xO_{12}$, as schematically shown in Figure 8. As reported in previous reports, $CaCu_3Ru_4O_{12}$ ($x = 0$) exhibits two kinds of electrical current paths (Figure 8a). The first one relies on the Ru-O network (I_{Ru}), while the second relies on Cu^{2+} (I_{Cu}). In the case of $x \neq 0$ (Figure 8b), the substitution of part of the Ru sites with Mn disrupts the conducting Ru-O network, and conduction electrons hop on to the Mn sites along their path (I_{Ru+Mn}). Hopping slows down the electrons, therefore I_{Ru+Mn} is smaller than I_{Ru}. However, the decrease in the total conduction current intensity is small as I_{Cu} is not affected by the presence of Mn.

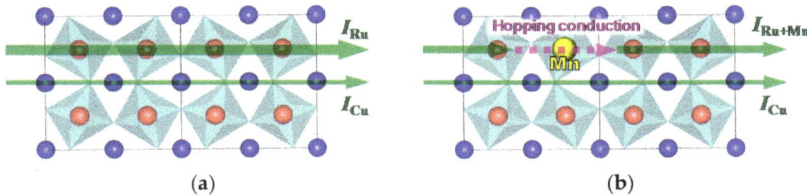

Figure 8. Schematic of $CaCu_3Ru_{4-x}Mn_xO_{12}$ in real space showing the different paths for electrical conduction when (**a**) $x = 0$, and (**b**) $x \neq 0$. For simplicity, Ca and O ions are omitted. ($CaCu_3Ru_{4-x}Mn_xO_{12}$ lattices were drawn using VESTA [26].)

Figure 9 shows the temperature dependence of the thermopower S (S-T curve) of the CuO(20 vol.%)-$CaCu_3Ru_{4-x}Mn_xO_{12}$ bulks. Similarly to the ρ-T curves, the S-T curves exhibit a systematic variation with x. The thermopower of all samples decreased with cooling, and the thermopower of the Mn-substituted samples was negative at low temperatures, while that of $CaCu_3Ru_4O_{12}$ was positive. A similar effect was observed for $Ca_{0.5}Sr_{0.5}RuO_3$ and $Ca_{0.5}Sr_{0.5}Ru_{0.5}Mn_{0.5}O_3$, whose thermopower between 10 and 280 K is respectively positive and negative between 10 and 280 K, explained by the fact that the Ru-Mn substitution changes the majority carrier from holes to electrons [28]. The temperature range in which the thermopower of the CuO(20 vol.%)-$CaCu_3Ru_{4-x}Mn_xO_{12}$ bulks is negative increased with increasing x, indicating that a gradual change of majority carrier took place in the CuO(20 vol.%)-CaCu$_3$Ru$_{4-x}$Mn$_x$O$_{12}$ bulks with increasing x. This Mn substitution effect is considered to be weaker in CaCu$_3$Ru$_{4-x}$Mn$_x$O$_{12}$ than in Ca$_{0.5}$Sr$_{0.5}$Ru$_{1-x}$Mn$_x$O$_3$ because of the A-site-based electrical conduction, which, as already mentioned, is not affected by B-site substitution.

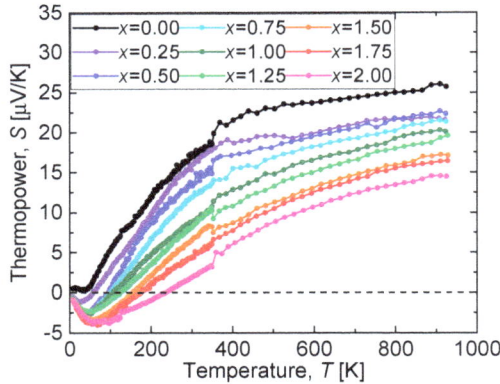

Figure 9. Temperature dependence of the thermopower of CuO(20 vol.%)-CaCu$_3$Ru$_{4-x}$Mn$_x$O$_{12}$ bulks.

It has been reported that the temperature dependence of the thermopower in the case of VRH conduction is of the form: $S = T^{1/2}$ [31]. Figure 10 shows the thermopower S of CaCu$_3$Ru$_2$Mn$_2$O$_{12}$ plotted against $T^{1/2}$. Although VRH conduction was observed below 215.89 K in the resistivity curve (215.89 K corresponds to $T^{1/2} = 14.7$ K$^{1/2}$, shown as a black dash line in Figure 10), the fitting line in Figure 10 (shown as a red solid line) corresponding to $|S| = aT^{1/2} + b$ indicates VRH conduction below $T^{1/2} = 7$ K$^{1/2}$ ($T = 49$ K, shown as a blue dot line). This discrepancy can be explained by the fact that the fitting equation for S has been derived from a model of an amorphous semiconductor with a single type of carrier, while both electrons and holes are electrical carriers in CaCu$_3$Ru$_{4-x}$Mn$_x$O$_{12}$. As S is strongly affected by the type of carrier, the red fitting line in Figure 10 fitted the $S(T^{1/2})$ curve only at very low temperatures, because holes lose mobility at these temperatures and electrons can be considered as the only charge carriers.

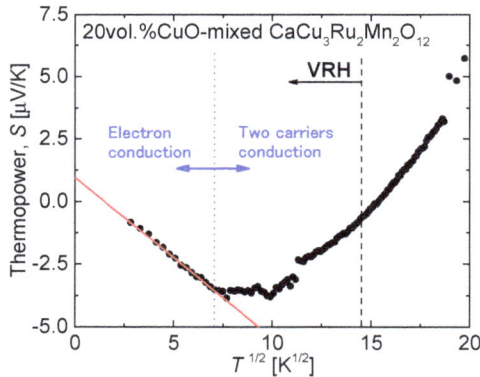

Figure 10. Thermopower of CuO(20 vol.%)-CaCu$_3$Ru$_2$Mn$_2$O$_{12}$ bulks plotted against $T^{1/2}$. The red solid line corresponds to the fitting equation $|S| = aT^{1/2} + b$.

Crystals **2017**, *7*, 213

4. Conclusions

We prepared CuO(20 vol.%)-CaCu$_3$Ru$_{4-x}$Mn$_x$O$_{12}$ bulks with various substitution amounts x, and investigated the influence of x on the electrical resistivity. Only the CuO and CaCu$_3$Ru$_{4-x}$Mn$_x$O$_{12}$ phases were detected by XRD in all samples, and a peak shift due to the substitution was confirmed. SEM observations and a calculation of the relative density showed an enhancement of the grain growth and sintering with increasing substitution. The resistivity increased with increasing x, but all samples maintained good conductivity with resistivity as low as a few mΩcm, even in the sample with $x = 2.00$. This phenomenon is explained by the existence of A-site (Cu^{2+}) conduction in CaCu$_3$Ru$_{4-x}$Mn$_x$O$_{12}$. The temperature dependence of the resistivity of CuO(20 vol.%)-CaCu$_3$Ru$_{4-x}$Mn$_x$O$_{12}$ bulks indicated a semiconducting behavior at low temperatures. The conduction mechanism at low temperatures was identified as variable range hopping conduction, where hopping occurs because the Ru conduction path is disrupted by the presence of Mn on Ru sites. The thermopower was found to be affected by the substitution as well. In particular, a sign inversion of the thermopower was observed in the substituted samples at low temperatures. This study therefore demonstrates that the partial substitution of Ru by Mn in CaCu$_3$Ru$_4$O$_{12}$ is an excellent strategy to reduce the material cost while maintaining good conductivity.

Acknowledgments: The authors thank Takara Hiroi (AIST, Japan) for her experimental assistance.

Author Contributions: A.T., I.T. and W.S. conceived and designed the experiments; A.T. performed the experiments; A.T., I.T. and W.S. analyzed the data; M.M., Y.K. and N.M. helped experiments and discussed about results; A.T. wrote the paper.

Conflicts of Interest: The authors declare no conflict of interest.

References and Notes

1. Marezio, M.; Dernier, P.D.; Chenavas, J.; Joubert, J.C. High pressure synthesis and crystal structure of NaMn$_7$O$_{12}$. *J. Solid State Chem.* **1973**, *6*, 16–20. [CrossRef]
2. Andres, K.; Graebner, J.E.; Ott, H.R. 4*f*-Virtual-Bound-State Formation in CeAl$_3$ at Low Temperatures. *Phys. Rev. Lett.* **1975**, *35*, 1779–1782. [CrossRef]
3. Bochu, B.; Deschizeaux, M.N.; Joubert, J.C.; Collomb, A.; Chenavas, J.; Marezio, M. Synthèse et caractérisation d'une série de titanates pérowskites isotypes de [CaCu$_3$](Mn$_4$)O$_{12}$. *J. Solid State Chem.* **1979**, *29*, 291–298. [CrossRef]
4. Subramanian, M.A.; Li, D.; Duan, N.; Reisner, B.A.; Sleight, A.W. High Dielectric Constant in *A*Cu$_3$Ti$_4$O$_{12}$ and *A*Cu$_3$Ti$_3$FeO$_{12}$ Phases. *J. Solid State Chem.* **2000**, *151*, 323–325. [CrossRef]
5. Subramanian, M.A.; Sleight, A.W. ACu$_3$Ti$_4$O$_{12}$ and ACu$_3$Ru$_4$O$_{12}$ perovskites: High dielectric constants and valence degeneracy. *Solid State Sci.* **2002**, *4*, 347–351. [CrossRef]
6. Calle, C.; Sánchez-Benítez, J.; Barbanson, F.; Nemes, N.; Fernández-Díaz, M.T.; Alonso, J.A. Transition from Pauli-paramagnetism to ferromagnetism in CaCu$_3$(Ru$_{4-x}$Mn$_x$)O$_{12}$ ($0 \leq x \leq 3$) perovskites. *J. Appl. Phys.* **2011**, *109*, 123914. [CrossRef]
7. Ramirez, A.P.; Subramanian, M.A.; Gardel, M.; Blumberg, G.; Li, D.; Vogt, T.; Shapiro, S.M. Giant dielectric constant response in a copper-titanate. *Solid State Commun.* **2000**, *115*, 217–220. [CrossRef]
8. Homes, C.C.; Vogt, T.; Shapiro, S.M.; Wakimoto, S.; Ramirez, A.P. Optical Response of High-Dielectric-Constant Perovskite-Related Oxide. *Science* **2001**, *293*, 673–676. [CrossRef] [PubMed]
9. Kondo, S.; Johnston, D.C.; Swenson, C.A.; Borsa, F.; Mahajan, A.V.; Miller, L.L.; Gu, T.; Goldman, A.I.; Maple, M.B.; Gajewski, D.A.; et al. LiV$_2$O$_4$: A Heavy Fermion Transition Metal Oxide. *Phys. Rev. Lett.* **1997**, *78*, 3729–3732. [CrossRef]
10. Long, Y.W.; Hayashi, N.; Saito, T.; Azuma, M.; Muranaka, S.; Shimakawa, Y. Temperature-induced A-B intersite charge transfer in an A-site-ordered LaCu$_3$Fe$_4$O$_{12}$ perovskite. *Nature* **2009**, *485*, 60–63. [CrossRef] [PubMed]
11. Kobayashi, W.; Terasaki, I.; Takeya, J.; Tsukada, I.; Ando, Y. A Novel Heavy-Fermion State in CaCu$_3$Ru$_4$O$_{12}$. *J. Phys. Soc. Jpn.* **2004**, *73*, 2373–2376. [CrossRef]

12. Hébert, S.; Daou, R.; Maignan, A. Thermopower in the quadruple perovskite ruthenates. *Phys. Rev. B* **2015**, *91*, 045106. [CrossRef]

13. Ebbinghaus, S.G.; Weidenkaff, A.; Cava, R.J. Structural Investigations of $ACu_3Ru_4O_{12}$ (A = Na, Ca, Sr, La, Nd)-Comparison between XRD-Rietveld and EXAFS Results. *J. Solid State Chem.* **2002**, *167*, 126–136. [CrossRef]

14. Fleischer, M.; Meixner, H. Fast gas sensors based on metal oxides which are stable at high temperatures. *Sens. Actuators B Chem.* **1997**, *43*, 1–10. [CrossRef]

15. Esch, H.; Huyberechts, G.; Mertens, R.; Maes, G.; Manca, J.; Ceuninck, W.; Schepper, L. The stability of Pt heater and temperature sensing elements for silicon integrated tin oxide gas sensors. *Sens. Actuators B Chem.* **2000**, *65*, 190–192. [CrossRef]

16. Izu, N.; Shin, W.; Murayama, N. Fast response of resistive-type oxygen gas sensors based on nano-sized ceria powder. *Sens. Actuators B Chem.* **2003**, *93*, 449–453. [CrossRef]

17. Korotcenkov, G.; Cho, B.K. Engineering approaches to improvement of conductometric gas sensor parameters. Part 2: Decrease of dissipated (consumable) power and improvement stability and reliability. *Sens. Actuators B Chem.* **2014**, *198*, 316–341. [CrossRef]

18. Matsuzaki, Y.; Yasuda, I. Electrochemical properties of a SOFC cathode in contact with a chromium-containing alloy separator. *Solid State Ion.* **2000**, *132*, 271–278. [CrossRef]

19. Suzuki, T.; Awano, M.; Jasinski, P.; Petrovsky, V.; Anderson, H.U. Composite (La, Sr)MnO$_3$–YSZ cathode for SOFC. *Solid State Ion.* **2006**, *177*, 2071–2074. [CrossRef]

20. Sumi, H.; Yamaguchi, T.; Hamamoto, K.; Suzuki, T.; Fujishiro, Y. High performance of $La_{0.6}Sr_{0.4}Co_{0.2}Fe_{0.8}O_3$–$Ce_{0.9}Gd_{0.1}O_{1.95}$ nanoparticulate cathode for intermediate temperature microtubular solid oxide fuel cells. *J. Power Sources* **2013**, *226*, 354–358. [CrossRef]

21. Jia, Q.X.; Wu, X.D.; Foltyn, S.R.; Findikoglu, A.T.; Tiwari, P.; Zheng, J.P.; Jow, T.R. Heteroepitaxial growth of highly conductive metal oxide RuO$_2$ thin films by pulsed laser deposition. *Appl. Phys. Lett.* **1995**, *67*, 1677–1679. [CrossRef]

22. Ryden, W.; Lawson, A.; Sartain, C. Electrical Transport Properties of IrO$_2$ and RuO$_2$. *Phys. Lett. B* **1970**, *1*, 1494–1500.

23. Pearsall, T.P.; Lee, C.A. Electronic transport in ReO$_3$: dc conductivity and Hall effect. *Phys. Rev. B* **1974**, *10*, 2190. [CrossRef]

24. Ullmann, H.; Trofimenko, N.; Tietz, F.; Stöver, D.; Ahmad-Khanlou, A. Correlation between thermal expansion and oxide ion transport in mixed conducting perovskite-type oxides for SOFC cathodes. *Solid State Ion.* **2000**, *138*, 79–90. [CrossRef]

25. Tsuruta, A.; Mikami, M.; Kinemuchi, Y.; Terasaki, I.; Murayama, N.; Shin, W. High electrical conductivity of composite ceramics consisting of insulating oxide and ordered perovskite conducting oxide. *Phys. Status Solidi A* **2017**, in press.

26. VESTA. Available online: http://jp-minerals.org/vesta/en/ (accessed on 1 April 2017).

27. Kolesnik, S.; Dabrowski, B.; Chmaissem, O. Structural and physical properties of SrMn$_{1-x}$Ru$_x$O$_3$ perovskites. *Phys. Rev. B* **2008**, *78*, 214425. [CrossRef]

28. Ohnishi, T.; Naito, M.; Mizusaki, S.; Nagata, Y.; Noro, Y. Transport and Thermoelectric Properties of the $Ca_{1-x}Sr_xRu_{1-y}Mn_yO_3$ System. *J. Electron. Mater.* **2011**, *40*, 915–919. [CrossRef]

29. Wang, L.; Hua, L.; Chen, L. F. First-principles investigation of the structural, magnetic and electronic properties of perovskite SrRu$_{1-x}$Mn$_x$O$_3$. *J. Phys. Condens. Matter* **2009**, *21*. [CrossRef] [PubMed]

30. Quitmann, C.; Andrich, D.; Jarchow, C.; Fleuster, M.; Beschoten, B.; Güntherodt, G.; Moshchalkov, V.V.; Mante, G.; Manzke, R. Scaling behavior at the insulator-metal transition in $Bi_2Sr_2(Ca_zR_{1-z})Cu_2O_{8+y}$ where R is a rare-earth element. *Phys. Rev. B* **1992**, *46*, 11813. [CrossRef]

31. Overhof, H. Thermopower Calculation for Variable Range Hopping-Application to α-Si. *Phys. Stat. Sol. B* **1975**, *67*, 709. [CrossRef]

© 2017 by the authors. Licensee MDPI, Basel, Switzerland. This article is an open access article distributed under the terms and conditions of the Creative Commons Attribution (CC BY) license (http://creativecommons.org/licenses/by/4.0/).

crystals

MDPI

Review

Ferroelectricity in Simple Binary Crystals

Akira Onodera * and Masaki Takesada

Department of Physics, Faculty of Science, Hokkaido University, Sapporo 060-0810, Japan;
mt@phys.sci.hokudai.ac.jp
* Correspondence: onodera@phys.sci.hokudai.ac.jp; Tel.: +81-11-706-2680

Academic Editor: Stevin Snellius Pramana
Received: 28 April 2017; Accepted: 23 May 2017; Published: 28 July 2017

Abstract: The origin of ferroelectricity in doped binary crystals, $Pb_{1-x}Ge_xTe$, $Cd_{1-x}Zn_xTe$, $Zn_{1-x}Li_xO$, and $Hf_{1-x}Zr_xO_2$ is discussed, while no binary ferroelectrics have been reported except for two crystals, HCl and HBr. The ferroelectricity is induced only in doped crystals, which shows an importance of electronic modification in chemical bonds by dopants. The phenomenological and microscopic treatments are given for the appearance of ferroelectric activity. The discovery of ferroelectricity in binary crystals such as ZnO and HfO_2 is of high interest in fundamental science and also in application for complementary metal–oxide semiconductor (CMOS) technology.

Keywords: ferroelectric; binary crystal; ZnO; HfO_2; mixed bond

1. Introduction

Ferroelectrics are expected as a key material for next-generation nonvolatile ferroelectric memories (FeRAM), piezoelectric actuators, high-k gate-materials for high-speed FET (field effect transistor), and optoelectronic devices [1–6]. Particularly, ferroelectric thin films such as perovskite PZT ($PbZr_{1-x}Ti_xO_3$) and Bi-layered perovskite SBT ($SrBi_2Ta_2O_9$) have been investigated extensively for FeRAM, because of their excellent dielectric properties, i.e., high dielectric constant and large spontaneous polarization. However, it is not so easy to fabricate good quality ferroelectric thin films on silicon substrate and integrate into devices overcoming degradation of ferroelectric properties due to the so-called size effect. Many ferroelectrics have crystal structures consisting of more than three atoms. For example, $BaTiO_3$, known as a typical ferroelectric with a perovskite structure, consists of three atoms. New materials with a simple structure are not only preferable for understanding the microscopic origin of ferroelectricity, but are also easy for integrating into modern ferroelectric devices. No ferroelectrics with two atoms have been reported except for two molecular crystals, HCl and HBr (Table 1) [7]. In 1969, Cochran developed the theory of lattice dynamics for alkali halide crystals such as NaCl and discussed the possibility of ferroelectricity, which revealed a real alkali halide crystal is not a ferroelectric, because a short-range restoring contribution is about twice as great as a long-range Coulomb contribution as discussed later [8–10]. However, it is important to point out that these two contributions are the same order for alkali halide crystals. We will show in this article that this balance of these two contributions may be modified by introducing some dopants, strain or defects in crystals.

Table 1. Phase diagram of HCl and HBr after Hoshino et al. [7].

	Phase III Ferroelectric	Phase II Paraelectric			Phase I Liquid
HCl	Orthorhombic-$Bb2_1m$	Cubic-$Fm3m$			Liquid
	Below 98 K	98~159 K			Above 159 K
HBr	Orthorhombic-$Bb2_1m$	Phase IIc Orthorhombic-$Bbcm$	Phase IIb Cubic	Phase IIa Cubic-$Fm3m$	Liquid
	Below 90 K	90~114 K	114~117 K	117~186 K	Above 186 K

The electronic ferroelectricity was found in wide-gap semiconductor ZnO by introducing a small amount of Li dopants, although pure ZnO does not show any evidence of ferroelectricity [11–15].

ZnO has a simple binary AB structure with high-symmetry (wurtzite structure). Besides ZnO, Ge-doped PbTe, a IV-VI narrow-gap semiconductor [16], and Zn-doped CdTe, a II-VI wide-gap semiconductor [17], have been investigated as materials of binary crystals accompanying ferroelectricity (Figure 1). Moreover, recent works showed that thin films of HfO_2-ZrO_2 systems exhibit ferroelectricity [18,19]. Hafnia (HfO_2) and Zirconia (ZrO_2) have been well studied as high-k dielectric materials in CMOS (complementary metal–oxide semiconductor) technology. Pure HfO_2 crystal is monoclinic with space group $P2_1/c$ at room temperature and atmospheric pressure. Only HfO_2 thin films doped with Si, Y, Al and Zr change the monoclinic crystal structure to a polar orthorhombic structure. The discovery of ferroelectricity in binary crystals such as ZnO and HfO_2 is of high interest in fundamental science and also in application fields. Our concern is to study the origin of this unexpected appearance of ferroelectricity in doped binary crystals. We will discuss why the ferroelectricity does not appear in pure systems but in doped crystals.

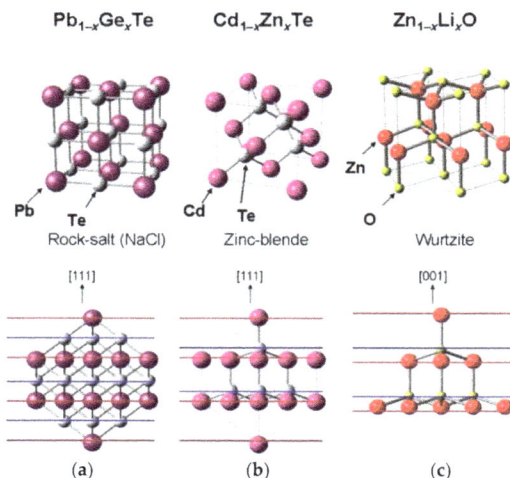

Figure 1. Crystal structures of (a) $Pb_{1-x}Ge_xTe$, (b) $Cd_{1-x}Zn_xTe$ and (c) $Zn_{1-x}Li_xO$. The lower figures are plots of cation (blue lines) and anion (red lines) layers along the polar rhombohedral [111] direction for $Pb_{1-x}Ge_xTe$ and $Cd_{1-x}Zn_xTe$, and polar [001] direction for $Zn_{1-x}Li_xO$ [15].

2. Ferroelectricity in Binary Semiconductors

Ferroelectrics, in general, have complicated crystal structures, which undergo a phase transition from a paraelectric high-temperature phase with decreasing temperature, breaking the symmetry of inversion. The dielectric constant (ε) and the spontaneous polarization (P_s) are characterized in the mean field approximation as

$$\varepsilon = C/(T - T_c) \tag{1}$$

$$P_s = P_o (T - T_c)^{1/2} \tag{2}$$

where T_c is a critical temperature. Ferroelectrics are generally classified into order-disorder, displacive and improper types, though multiferroic materials have been reported recently. According to these types, they show the following characteristic dielectric properties as summarized in Table 2 [20].

Table 2. Typical values of the Curie-Weiss constant (*C*), the dielectric constant around T_c (ε_{max}), and the spontaneous polarization (P_s) according to the type of ferroelectrics.

Type of Ferroelectrics	*C* [K]	ε_{max} at T_c	P_s [μC/cm^2]	Example
Order-disorder	$1\sim3 \times 10^3$	10^3	$3\sim5$	TGS [§]
Displacive	10^5	10^4	$10\sim30$	BaTiO$_3$
Improper	10	10	Small	ACS [#]
Electronic	-	21	0.9	ZnO

[§] TGS (NH$_3$CH$_2$COOH)$_3$H$_2$SO$_4$), [#] ACS ((NH$_4$)$_2$Cd$_2$(SO$_4$)$_3$).

Binary ferroelectric crystals show different but common dielectric behavior from those of the usual ferroelectrics described above; a small dielectric anomaly at T_c and relatively large P_s. We will review the ferroelectric properties of the ferroelectric binary semiconductors, PbTe, CdTe and ZnO, and a high-*k* dielectric HfO$_2$, briefly in this section. More detailed discussion should be referred in a monograph [15].

2.1. Narrow-Gap Ferroelectric Semiconductor PbTe

The PbTe-GeTe system has been investigated extensively about its ferroelectricity among IV-VI semiconductors [16], which has a rock-salt type structure (*Fm3m*, *a* = 6.46 Å) at room temperature. The energy gap (E_g) is 0.3 eV, which is comparable to the Lorentz field ($4\pi/3$)P. The ferroelectricity is observed in solid solution Pb$_{1-x}$Ge$_x$Te. The stacking Pb^{2+} cation and Te^{2-} anion layers dimerize along the rhombohedral [111] direction, as shown in Figure 1. The crystal changes to a rhombohedral structure (*R3m*) which allows it to exhibit ferroelectric activity Pb$_{1-x}$Ge$_x$Te with *x* = 0.003 shows a large dielectric anomaly at *T* = 100 K.

The large dielectric anomaly and the existence of the soft mode suggest the ferroelectric activity.

2.2. Wide-Gap Ferroelectric Semiconductor CdTe

Cd$_{1-x}$Zn$_x$Te is a II-VI wide-gap semiconductor with E_g = 1.53 eV. Weil et al. discovered the ferroelectric activity in Cd$_{1-x}$Zn$_x$Te, as shown in Figure 2 [17,21]. The cubic zinc-blende structure (space group $F\overline{4}3m$, *a* = 6.486 Å) of pure CdTe crystal changes to a rhombohedral one (*R3m*, *a* = 6.401 Å, α = 89.94°) in Cd$_{1-x}$Zn$_x$Te, as shown in Figure 1b. The dielectric anomaly at T_c (393 K) is smaller by two orders than that of typical ferroelectric BaTiO$_3$ (~14,000). The spontaneous polarization is about 5 μC/cm^2 along the rhombohedral [111] direction. Doped Zn ions locate at *off-center* positions [22] which cause rhombohedral distortion of about 0.01 Å in Cd$_{1-x}$Zn$_x$Te [22].

No soft mode has been observed in Cd$_{1-x}$Zn$_x$Te, and the dielectric anomaly is small. Although the behavior of *off-center ions* plays an important role in this ferroelectricity like Pb$_{1-x}$Ge$_x$Te, the occurrence of phase transition seems to be driven in a different way from that of Pb$_{1-x}$Ge$_x$Te.

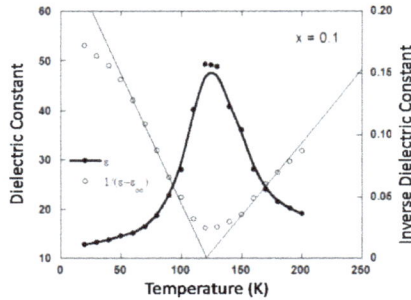

Figure 2. Dependence of dielectric constant and inverse dielectric constant in $Cd_{1-x}Zn_xTe$ ($x = 0.1$) on temperature [17]. Reprinted figure with permission from R. Weil, R. Nkum, E. Muranevich, and L. Benguigui, Physical Review Letters, 62, 2744, 1989. Copyright (1989) by the American Physical Society.

2.3. II-VI Wide-Gap Semiconductor ZnO

Zinc Oxide (ZnO), a II-VI wide-gap semiconductor, is a well-studied electronic material with a large piezoelectric constant [23–27]. ZnO has been studied as materials for solar cells, transparent conductors and blue lasers [28,29]. This crystal has a wurtzite structure ($P6_3mc$) (Figure 3). This space group is non-centrosymmetric and is allowed to exhibit ferroelectricity, although no *D-E* loop has been observed until melting point. Introduction of a small amount of Li-dopants results in the ferroelectricity.

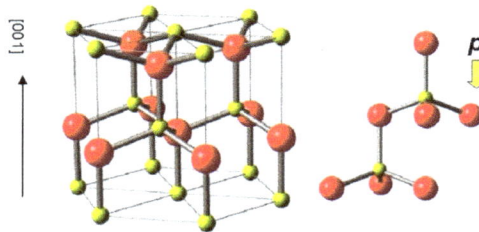

Figure 3. Crystal structure of ZnO. The observed polarization (*p*) is shown by a yellow arrow [15].

A dielectric anomaly in $Zn_{1-x}Li_xO$ ($x = 0.09$) was found at 470 K (T_c) (Figure 4), though pure ZnO shows no anomaly from 20 K to 700 K. The small dielectric anomaly ($\varepsilon_{max} = 21$) is the same order with $Cd_{1-x}Zn_xTe$ ($\varepsilon_{max} = 50$).

The spontaneous polarization is 0.9 $\mu C/cm^2$ [30,31]. The phase diagram between T_c and x is shown in Figure 5, which reminds us of a phase diagram of quantum ferroelectrics such as $KTa_{1-x}Nb_xO_3$ and $Sr_{1-x}Ca_xTiO_3$. Raman scattering measurements showed no soft modes [32,33].

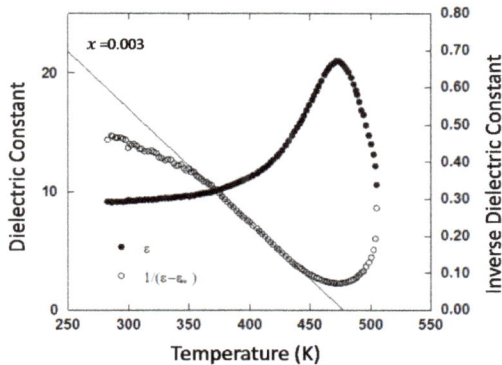

Figure 4. Temperature dependence of the dielectric constant of $Zn_{1-x}Li_xO$ (x = 0.09) [15].

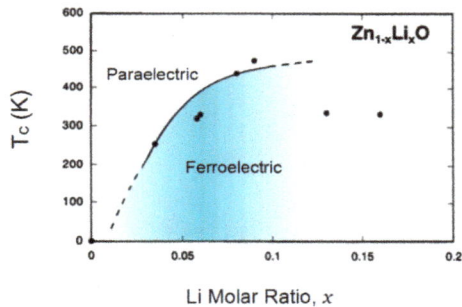

Figure 5. Phase diagram of the ferroelectric transition temperature (T_c) vs. Li molar ratio (x) in $Zn_{1-x}Li_xO$ [15].

2.4. High-k Materials HfO_2 and ZrO_2

HfO_2 and ZrO_2 are well-known high-temperature dielectrics. The thin films of HfO_2 and ZrO_2 have been studied extensively as a high-k gate dielectric film in CMOS technology. The crystal structure is monoclinic with space group $P2_1/c$ at room temperature and atmospheric pressure, which transforms to a tetragonal structure ($P4_2/nmc$) at ~1990 K, and then to a cubic Auorite structure ($Fm3m$) at ~2870 K. Other two orthorhombic phases have been reported under high pressure near 4 and 14 GPa [34]. Figure 6 shows a *P-T* phase diagram of HfO_2.

The monoclinic $P2_1/c$ structure is centrosymmetric, which does not show any ferroelectric activity. Thin films of HfO_2 undergo a phase transition to a noncentrosymmetric orthorhombic structure, breaking the symmetry of inversion when the films are doped with Si, Y, Al, or Zr [18,19,35–38]. The X-ray diffraction patterns of HfO_2, $Hf_{0.5}Zr_{0.5}O_2$ and ZrO_2 thin films are shown in Figure 7 [38]. Four space groups ($Pmn2_1$, $Pca2_1$, $Pbca$, and $Pbcm$) are proposed for this orthorhombic ferroelectric phase. Among these space groups, it is considered that $Pca2_1$ is the most probable. The sequence of phase transitions and crystal structures are shown in Figure 8. The ferroelectricity was confirmed by *D-E* hysteresis measurements, which revealed P_s of 16 $\mu C/cm^2$ as shown in Figure 9. The Curie temperature (T_c) was estimated to be about 623 K [39,40].

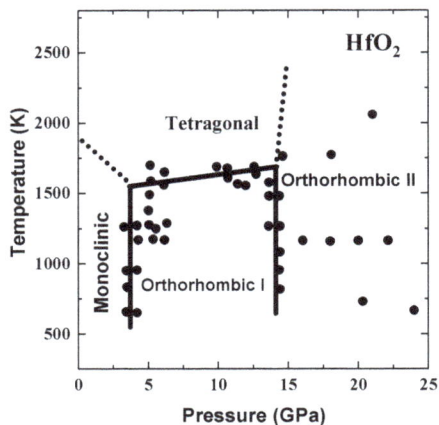

Figure 6. *P–T* phase diagram of HfO_2 [34].

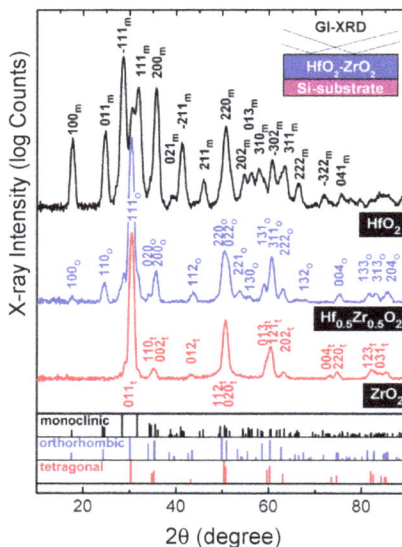

Figure 7. X-ray diffraction patterns of HfO_2, $Hf_{0.5}Zr_{0.5}O_2$ and ZrO_2 thin films on silicon substrates. The Bragg reflections are assigned as *h k l* with suffixes *m*, *o* and *t*, which indicate monoclinic, orthorhombic and tetragonal lattices, respectively [38]. Reprinted with permission from Johannes Müller, Tim S. Böscke, Uwe Schröder, et al., Ferroelectricity in Simple Binary ZrO_2 and HfO_2, Nano Letters, 2012, 12, 4318–4323. Copyright (2012) American Chemical Society.

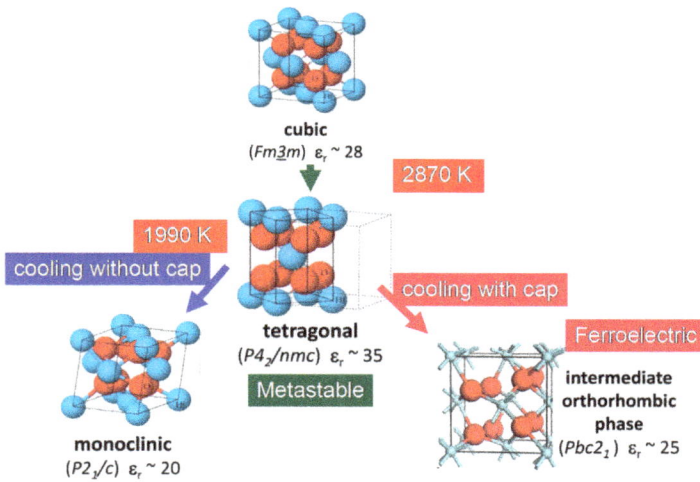

Figure 8. Sequence of phase transitions and crystal structures of HfO_2.

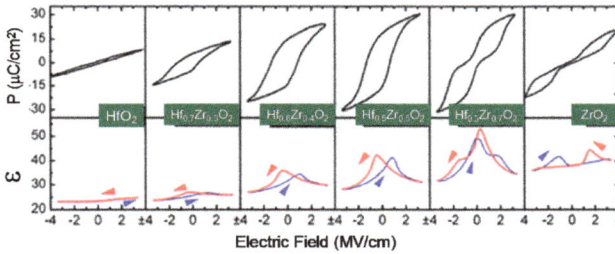

Figure 9. *P-E* hysteresis loops and dielectric constant (ε) in thin films of HfO_2-ZrO_2 system [38]. Reprinted with permission from Johannes Müller, Tim S. Böscke, Uwe Schröder, et al., Ferroelectricity in Simple Binary ZrO_2 and HfO_2, Nano Letters, 2012, 12, 4318–4323. Copyright (2012) American Chemical Society.

3. Phenomenological Treatment for the Appearance of Ferroelectricity

Firstly following the Landau theory, we consider the free energy (F) for the paraelectric phase of binary crystals in terms of polarization P as

$$F = 1/2\alpha P^2 + 1/4\beta P^4 + \dots, \tag{3}$$

where α, and β are coefficients. In general, the only coefficient α depends on temperature as

$$\alpha = \alpha_o (T - T_o), \alpha_o > 0 \tag{4}$$

In the case for paraelectric dielectrics, the critical temperature T_o is considered to be lower than 0 K, because of the absence of phase transitions above 0 K. When some dopants are introduced to crystals, some structural changes due to a difference in atomic radii and bonding electrons are modified by dopants. These may induce some local changes in electronic distribution in crystals. These extra contributions could be added to the above free energy as

$$F = 1/2\alpha P^2 + 1/4\beta P^4 + g\eta P + 1/2\alpha'\eta^2 \dots , \qquad (5)$$

where the last term $1/2\alpha'\eta^2$ is the contribution by dopants, and $g\eta P$ is an interaction term between the host crystal and extrinsic dopants. From the stability condition $\partial F/\partial\eta = 0$,

$$\eta = -(g/\alpha')P, \qquad (6)$$

The above free energy F can be rewritten as

$$F = 1/2(\alpha - g^2/\alpha')P^2 + 1/4\beta P^4 \qquad (7)$$

Using Equation (4), we get

$$(\alpha - g^2/\alpha') = \alpha_0 (T - T_c) \qquad (8)$$

where

$$T_c = T_o + g^2/\alpha_0\alpha' \qquad (9)$$

As the critical temperature T_c increases when the coefficient α' is positive, it should be possible to undergo a ferroelectric phase transition. T_c shows a rapid increase in the order of 10^2 K for large g and small α' in the case of $Pb_{1-x}Ge_xTe$, $Cd_{1-x}Zn_xTe$, and $Zn_{1-x}Li_xO$. The appearance of ferroelectricity is realized in the doped binary crystals, while any phase transition does not occur in pure crystals.

4. Microscopic Consideration after Cochran's Lattice Dynamical Theory

Although the above phenomenological theory can explain well the appearance of ferroelectricity, it is not so easy for us to understand what kind of phenomena occurs in real crystals. For binary crystals such as NaCl, Cochran proposed a lattice dynamical theory to elucidate the origin of ferroelectric phase transition based on the shell model [9,10]. We review simply Cochran's soft mode theory for ferroelectrics at first.

Cochran calculated the frequencies of transverse and optic phonon modes in a diatomic cubic crystal using a shell model such as NaCl, as illustrated in Figure 10. The shell is originated from some local lattice deformation, electronic overlap forces, or covalency in chemical bonds. The core of the negative ion (charge Xe, mass m_2) will interact through an outer shell (charge Ye, mass ~0) with a force constant k. The positive ion (charge Ze, mass m_1) interacts with the shell by a short range force through a force constant R_0. The displacements are denoted as u_1, v_2 and u_2 for the positive ion, shell and negative core, respectively.

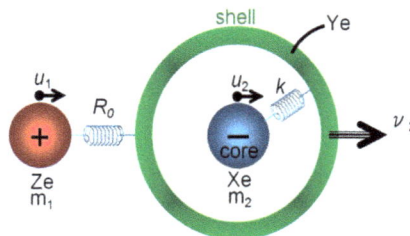

Figure 10. Schematic diagram of a shell model for a diatomic crystal.

The frequencies of the transverse and longitudinal optic modes, ω_{TO} and ω_{LO} are calculated as

$$\mu\omega_{TO}^2 = R_o' - \frac{4\pi(\varepsilon_\infty + 2)(Z'e)^2}{9V} \tag{10}$$

$$\mu\omega_{LO}^2 = R_o' + \frac{8\pi(\varepsilon_\infty + 2)(Z'e)^2}{9V\varepsilon_\infty} \tag{11}$$

where

$$R_o' = \frac{kR_0}{k + R_0} < R_o \tag{12}$$

$$Z' = Z + \frac{YR_0}{k + R_0} < Z \tag{13}$$

$$\mu = \frac{m_1 m_2}{m_1 + m_2} \tag{14}$$

ε_∞ and V are a high-frequency dielectric constant and the unit cell volume. Following the Lyddane-Sachs-Teller (LST) relation, the dielectric constant ε is given as

$$\frac{\varepsilon}{\varepsilon_\infty} = \frac{\omega_{LO}^2}{\omega_{TO}^2} \tag{15}$$

The ferroelectric phase is realized if $\omega_{TO} = 0$ in Equation (10), because ferroelectric phase transitions generally accompany with a divergence of dielectric constant. The ferroelectricity is induced from the delicate balance between the short-range term R_0' and the second dipolar Coulomb term in the right side of Equation (10).

Cochran showed that a real alkali halide crystal is not a ferroelectric, because R_0' is about twice as great as the other while these two contributions are the same order for real alkali halide crystals. This is one reason why the ferroelectricity has been found only in HCl and HBr. Dopants will change local electronic distribution of chemical bond, i.e., the nature of covalency, particularly in mixed bonded crystals. Dopants also force it to displace to *off-centered* positions and induce local structural distortions. Particularly, the chemical bonds may be affected sensitively by dopants in ZnO, CdTe and HfO_2 where the degree of covalency (or iconicity) is nearly half, as shown in Table 3. The balance between the restoring force R_0' and the dipolar Coulomb force can be modified by dopants, defects or strain.

Table 3. Dielectric constant and fractional degree of iconicity after J.C. Phillips, Bonds and Bands in Semiconductors [41].

Materials	ε at R.T.	Fractional Degree of Ionicity
NaCl	5.6	0.94
MgO	10.0	0.84
ZnO	8.8	0.62
CdTe	7.1	0.67
HfO_2	25	0.8
ZrO_2	30~46	0.8
TiO_2	170	0.6
Si	3.6	0

5. Discussion

5.1. Electronic Ferroelectricity in ZnO; Effect of Dopants

The replacement of host Zn ions by substitutional Li ions plays a primary role for the appearance of ferroelectricity in ZnO. To clarify the effect of dopants, structural size-mismatch and electronic models are studied: the introduction of small Be^{2+} ions (ionic radius 0.3 Å) should be effective than

Li$^+$ (ionic radius 0.6 Å) and Mg^{2+} (ionic radius 0.65 Å) ions if the ionic size-mismatch is important for ferroelectricity, while Mg^{2+} ions ($1s^2 2s^2 2p^6$) should play a different role from the isoelectronic Li$^+$ and Be^{2+} ions ($1s^2$) if the electronic configuration is important.

The series of dielectric measurements show that the introduction of Mg^{2+} ions suppresses T_c [42]. The appearance of ferroelectricity is primarily due to electronic origin.

5.2. Structural Modification by Dopants

The electronic distribution, especially the nature of *d-p* hybridization of paraelectric pure ZnO and ferroelectric Zn$_{1-x}$Li$_x$O at 19 K, was measured directly by X-ray diffraction. The main difference is observed in electronic distribution around Zn ion, as shown in Figure 11.

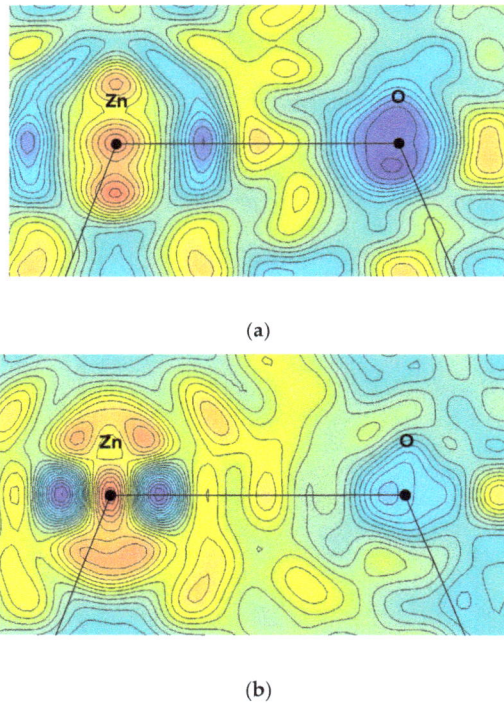

(a)

(b)

Figure 11. The difference Fourier maps of (**a**) paraelectric ZnO and (**b**) ferroelectric Zn$_{1-x}$Li$_x$O at 19 K in the (110) plane with a contour increment of 0.2 $e/$Å^3. The horizontal straight lines from Zn to O ions are the [001] direction. Bluish cold color means negative charge density and reddish warm region is positive charge density.

The negative distribution is observed around the Zn atom in Zn$_{1-x}$Li$_x$O, whose shape corresponds to Zn-3$d_z{}^2$-orbital. This evidence shows that the Zn 3d-electrons disappear around Zn position in the doped ZnO.

Most crystals have a fraction of covalent and ionic bonding components. ZnO is bonded half by ionic and half by covalent forces, of which delicate balance is slightly changed by Li-dopants without *d-electrons*. Pure ZnO has large dipole moment [11], while this host dipole could not be reversible by an electric field. In Zn$_{1-x}$Li$_x$O, the dipole is reduced a little bit by the local electronic deformation along the polar (0 0 1) direction; this is, in other words, an introduction of *negative dipoles* in the host lattice. These *negative dipoles* are responsible for an electric field and behave as *"hole dipoles"* which are similar to "*hole electrons*" in *n*-type semiconductors. We should call this type of ferroelectrics as "*n-type ferroelectrics*", while usual ferroelectrics are "*p-type ferroelectrics*".

In the case of $Hf_{1-x}Zr_xO_2$, the ionic radii (0.83 Å for Hf^{4+} and 0.84 Å for Zr^{4+}) are almost the same. The main difference is an electronic structure, Hf ($-4f^{14}$) and Zr ($-4p^6$). As the high dielectric constants in these crystals ($\varepsilon = 25$ for HfO_2, $\varepsilon = 30\sim46$ for ZrO_2) mean the large splitting of ω_{TO} and ω_{LO} phonon modes, a ω_{TO} phonon mode is much lower compared with a ω_{LO} (LST relation, Equation (10)). In this sense, HfO_2 and ZrO_2 are incipient ferroelectrics as $SrTiO_3$. The electronic distribution in the Hf-O bond is perturbed by Zr dopants without *f*-electrons. This modification may stabilize an orthorhombic polar structure and results in the appearance of ferroelectricity in $Hf_{1-x}Zr_xO_2$.

The above microscopic consideration gives us one perspective to understand the origin of ferroelectricity in doped binary crystals. If the condition that ω_{TO} approaches to zero at T_c holds exactly, soft mode should be detectable. However, no soft mode has been observed in $Cd_{1-x}Zn_xTe$, $Zn_{1-x}Li_xO$ and $Hf_{1-x}Zr_xO_2$, except for $Pb_{1-x}Ge_xTe$. This evidence suggests that the mechanism of the ferroelectric phase transition is not so simple in real crystals which are partially covalent and partially ionic. The small dielectric anomaly in $Cd_{1-x}Zn_xTe$, $Zn_{1-x}Li_xO$ and $Hf_{1-x}Zr_xO_2$ reminds us of an improper type ferroelectrics, rather than the order-disorder-type and displacive-type of ferroelectrics. This problem has been left for our future studies.

6. Another Possible Ferroelectric TiO$_2$

Rutile TiO_2 is studied well by various techniques [43]. The structure is tetragonal with space group $P4_2/mnm$ ($a = 4.593659$ (18) and $c = 2.958682$ (8) Å at 298 K, $Z = 2$) (Figure 12) [44]. Besides rutile, TiO_2 admits another two polymorphic forms in nature, i.e., anatase ($I4_1/amd$, $a = 3.7845$, $c = 9.5143$ Å, $Z = 4$) and brookite (*Pbca*, $a = 5.4558$, $b = 9.1819$, $c = 5.1429$ Å, $Z = 8$). Rutile TiO_2 is the most common of the three polymorphic forms. Under high pressure, TiO_2 undergoes a series of structural phase transitions.

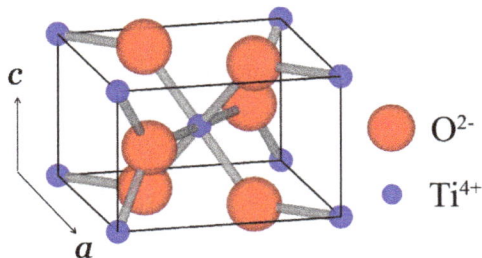

Figure 12. Crystal structure of Rutile TiO$_2$. Ti (blue) and O (red).

Rutile TiO_2 has large refractive indices ($n_c = 2.903$, $n_a = 2.616$) and large static dielectric constants ($\varepsilon_c = 170$, $\varepsilon_a = 86$) at room temperature.

Parker measured the dielectric constant ε of TiO_2 which increases with decreasing temperature, but does not show any anomaly from 1.6 to 1060 K, as shown in Figure 13 [45,46]. Pure TiO_2 does not show any ferroelectric or antiferroelectric activity. The dielectric constant shows a plateau at low temperatures around 0 K, which reminds us of dielectric behavior in quantum paraelectrics, such as $SrTiO_3$.

Although there had been a long-running discussion concerning the covalency of the bonding in rutile, Gronschorek [47] and Sakata et al. [48] concluded that Ti-O bonding in rutile is largely covalent, as shown in Figure 14. If we can modify the nature of bonding by some dopants or stress, it may be possible to expect the appearance of ferroelectricity.

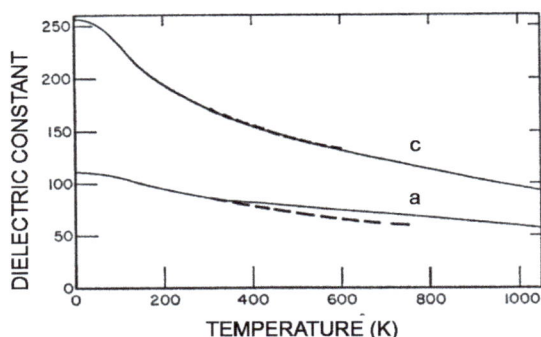

Figure 13. Temperature dependence of dielectric constant of rutile TiO_2 along the *a*- and *c*-directions. The solid lines are after Parker [45] and the dashed lines after von Hippel [46].

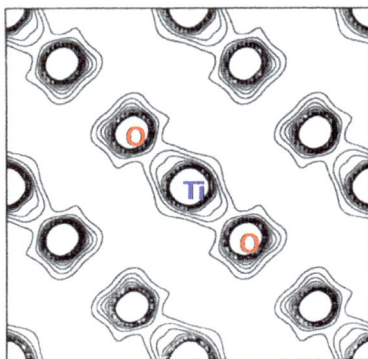

Figure 14. Electron density map of rutile TiO_2 on the (0 0 2) plane. Charge of 0.4 e $Å^{-3}$ is observed in Ti-O bond.

Recently, Montanari and Harrison proposed by density functional calculations that ferroelectric instability can be possible in rutile TiO_2 by applying a negative isotropic pressure [49]. The TO A2u mode, which is the *c*-axis ferroelectric, vanishes at -4 GPa, thereby leading to a crystal instability.

Similar calculations on binary oxides such as BaO, CaO, MgO, EuO, SnO_2 have been reported recently [50,51], although these crystals are primarily ionic (the degree of ionicity is ~0.8). They showed that ferroelectricity can be induced even in simple alkaline-earth-metal binary oxides by using appropriate epitaxial strains in thin films or in nano-particles.

7. Conclusions

We discussed the origin of ferroelectricity in doped binary crystals, $Pb_{1-x}Ge_xTe$, $Cd_{1-x}Zn_xTe$, $Zn_{1-x}Li_xO$, and $Hf_{1-x}Zr_xO_2$ on the basis of phenomenological and lattice dynamical treatments, while no ferroelectrics have been reported in pure binary crystals except for HCl and HBr. The delicate balance of the short-range restoring force and the long-range dipolar Coulomb force is tuned by dopants, particularly in binary crystals which are half covalent and half ionic. The modification of the electronic distribution in the chemical bond results in the local structural distortion, which may stabilize a non-centrosymmetric polar structure from a paraelectric structure with high-symmetry.

The discovery of ferroelectricity in doped binary crystals shows us a richness of structural science. Moreover, the ferroelectric HfO_2 is expected to be a promising candidate for FeRAM and high-speed

FET. Ferroelectrics are a group of materials sensitive to small structural changes. We must take into account the electronic contribution in the case of simple binary crystals discussed here. Further precise structural and theoretical studies should be necessary to clarify the possibility of ferroelectricity in other doped binary crystals.

Acknowledgments: This work was partially supported by Grant-in-Aid for Scientific Research (C) from JSPS No. 26400306 and a research granted from The Murata Science Foundation.

Conflicts of Interest: The authors declare no conflict of interest.

References

1. Auciello, O.; Scott, J.F.; Ramesh, R. Ultrahigh-Intensity Lasers: Physics of the Extreme on a Tabletop. *Phys. Today* **1998**, *51*, 22. [CrossRef]
2. Scott, J.F. The physics of ferroelectric ceramic thin films for memory applications. *Ferroelectr. Rev.* **1998**, *1*, 1. [CrossRef]
3. Haertling, G.H. Ferroelectric Ceramics: History and Technology. *J. Am. Ceram. Soc.* **1999**, *82*, 797–818. [CrossRef]
4. Scott, J.F.; Paz de Araujo, C.A. Ferroelectric memories. *Science* **1989**, *246*, 1400–1405. [CrossRef] [PubMed]
5. Uchino, K. Ceramic actuators: Principles and applications. *MRS Bull.* **1993**, *18*, 42–48. [CrossRef]
6. Wemple, S.H.; DiDomenico, M., Jr. Oxygen-Octahedra Ferroelectrics. II. Electro-optical and Nonlinear-Optical Device Applications. *J. Appl. Phys.* **1969**, *40*, 735. [CrossRef]
7. Hoshino, S.; Shimaoka, S.; Niimura, K. Ferroelectricity in Solid Hydrogen Halides. *Phys. Rev. Lett.* **1967**, *19*, 1286. [CrossRef]
8. Cochran, W. Crystal Stability and the Theory of Ferroelectricity. *Phys. Rev. Lett.* **1959**, *3*, 412. [CrossRef]
9. Cochran, W. Crystal stability and the theory of ferroelectricity. *Adv. Phys.* **1960**, *9*, 387. [CrossRef]
10. Cochran, W. Crystal stability and the theory of ferroelectricity part II. Piezoelectric crystals. *Adv. Phys.* **1961**, *10*, 401. [CrossRef]
11. Corso, D.A.; Posternak, M.; Resta, R.; Baldereshi, A. Ab initio study of piezoelectricity and spontaneous polarization in ZnO. *Phys. Rev.* **1994**, *B50*, 10715. [CrossRef]
12. Tamaki, N.; Onodera, A.; Sawada, T.; Yamashita, H. Measurements of D-E Hysteresis Loop and Ferroelectric Activity in Piezoelectric Li-doped ZnO. *J. Korean Phys.* **1996**, *29*, 668.
13. Onodera, A.; Tamaki, N.; Kawamura, Y.; Sawada, T.; Yamashita, H. Dielectric Activity and Ferroelectricity in Piezoelectric Semiconductor Li-Doped ZnO. *Jpn. J. Appl. Phys.* **1996**, *35*, 5160. [CrossRef]
14. Onodera, A.; Tamaki, N.; Jin, K.; Yamashita, H. Ferroelectric Properties in Piezoelectric Semiconductor Zn$_{1-x}$M$_x$O (M=Li, Mg). *Jpn. J. Appl. Phys.* **1997**, *36*, 6008. [CrossRef]
15. Onodera, A.; Takesada, M. Electronic Ferroelectricity in II-VI Semiconductor ZnO. In *Advances in Ferroelectrics*; INTECH: Rijeka, Croatia, 2012.
16. Bilz, H.; Bussmann-Holder, A.; Jantsch, W.; Vogel, P. *Dynamical Properties of IV-VI Compounds*; Springer: Berlin, Germany, 1983.
17. Weil, R.; Nkum, R.; Muranevich, E.; Benguigui, L. Ferroelectricity in zinc cadmium telluride. *Phys. Rev. Lett.* **1989**, *62*, 2744. [CrossRef] [PubMed]
18. Böscke, T.S.; Müller, J.; Brauhaus, D.; Schröder, U.; Böttger, U. Ferroelectricity in hafnium oxide thin films. *Appl. Phys. Lett.* **2011**, *99*, 102903. [CrossRef]
19. Böscke, T.S.; Teichert, S.; Brauhaus, D.; Müller, J.; Schröder, U.; Böttger, U.; Mikolajick, T. Phase transitions in ferroelectric silicon doped hafnium oxide. *Appl. Phys. Lett.* **2011**, *99*, 112904. [CrossRef]
20. Mitsui, T.; Tatsuzaki, I.; Nakamura, E. *An Introduction to the Physics of Ferroelectrics*; Gordon and Breach: New York, NY, USA, 1976; p. 202.
21. Benguigui, L.; Weil, R.; Muranevich, E.; Chack, A.; Fredj, E. Ferroelectric properties of Cd$_{1-x}$Zn$_x$Te solid solutions. *J. Appl. Phys.* **1993**, *74*, 513. [CrossRef]
22. Terauchi, H.; Yoneda, Y.; Kasatani, H.; Sakaue, K.; Koshiba, T.; Murakami, S.; Kuroiwa, Y.; Noda, Y.; Sugai, S.; Nakashima, S.; et al. Ferroelectric behaviors in semiconductive Cd$_{1-x}$Zn$_x$Te crystals. *Jpn. J. Appl. Phys.* **1993**, *32*, 728. [CrossRef]

23. Klingshirn, C.F.; Meyer, B.K.; Waag, A.; Hoffmann, A.; Geurts, J. *Zinc Oxide From Fundamental Properties Towards Novel Applications*; Springer: Berlin, Germany, 2010.

24. Yao, T. (Ed.) *ZnO Its Most Up-to-Date Technology and Application, Perspectives*; CMC Books: Tokyo, Japan, 2007. (In Japanese)

25. Heiland, G.; Mollwo, E.; Stockmann, F. Electronic Processes in Zinc Oxide. *Solid State Phys.* **1959**, *8*, 191–323. [CrossRef]

26. Campbell, C. *Surface Acoustic Wave Devices and Their Signal Processing Application*; Academic Press: San Diego, CA, USA, 1989.

27. Hirshwald, W.; Bonasewicz, P.; Ernst, L.; Grade, M.; Hofmann, D.; Krebs, S.; Littbarski, R.; Neumann, G.; Grunze, M.; Kolb, D.; et al. *Current Topics in Materials Science*; Kaldis, E., Ed.; North-Holland: Amsterdam, The Netherlands, 1981; Volume 7, p. 148.

28. Tsukazaki, A.; Ohtomo, A.; Onuma, T.; Ohtani, M.; Makino, T.; Sumiya, M.; Ohtani, K.; Chichibu, S.F.; Fuke, S.; Segawa, Y.; et al. Repeated temperature modulation epitaxy for p-type doping and light-emitting diode based on ZnO. *Nat. Mater.* **2005**, *4*, 42–46. [CrossRef]

29. Joseph, M.; Tabata, H.; Kawai, T. p-Type Electrical Conduction in ZnO Thin Films by Ga and N Codoping. *Jpn. J. Appl. Phys.* **1999**, *38*, L1205. [CrossRef]

30. Onodera, A.; Yoshio, K.; Satoh, H.; Yamashita, H.; Sakagami, N. Li-Substitution Effect and Ferroelectric Properties in Piezoelectric Semiconductor ZnO. *Jpn. J. Appl. Phys.* **1998**, *37*, 5315. [CrossRef]

31. Onodera, A.; Tamaki, N.; Satoh, H.; Yamashita, H.; Sakai, A. Novel ferroelectricity in piezoelectric ZnO by Li-substitution. In *Dielectric Ceramic Materials: Ceramic Transactions*; Nair, K.M., Bhalla, A.S., Eds.; American Ceramic Society: Westerville, OH, USA, 1999; Volume 100, pp. 77–94.

32. Islam, E.; Sakai, A.; Onodera, A. Optical Phonons in Ferroelectric-Semiconductor Zn 0.8Li 0.2O Single Crystal Studied by Micro-Raman Scattering. *J. Phys. Soc. Jpn.* **2001**, *70*, 576. [CrossRef]

33. Kagami, D.; Takesada, M.; Onodera, A.; Satoh, H. Photoinduced Effect in Li-doped ZnO studied by Raman Scattering. *J. Korean Phys.* **2011**, *59*, 2532. [CrossRef]

34. Ohtaka, O.; Fukui, H.; Kunisada, T.; Fujisawa, T.; Funakoshi, K.; Utsumi, W.; Irifune, T.; Kuroda, K.; Kikegawa, T. Phase relations and volume changes of hafnia under high pressure and high temperature. *J. Am. Ceram. Soc.* **2001**, *84*, 1369–1373. [CrossRef]

35. Müller, J.; Schröder, U.; Böscke, T.S.; Müller, I.; Böttger, U.; Wilde, L.; Sundqvist, J.; Lemberger, M.; Kücher, P.; Mikolajick, T.; et al. Ferroelectricity in yttrium-doped hafnium oxide. *J. Appl. Phys.* **2011**, *110*, 114113. [CrossRef]

36. Müller, S.; Müller, J.; Singh, A.; Riedel, S.; Sundqvist, J.; Schröder, U.; Mikolajick, T. Incipient Ferroelectricity in Al-Doped HfO₂ Thin Films. *Adv. Funct. Mater.* **2012**, *22*, 2412–2417. [CrossRef]

37. Müller, J.; Böscke, T.S.; Brauhaus, D.; Schröder, U.; Böttger, U.; Sundqvist, J.; Kücher, P.; Mikolajick, T.; Frey, L. Ferroelectric Zr₀.₅Hf₀.₅O₂ thin films for nonvolatile memory applications. *Appl. Phys. Lett.* **2011**, *99*, 112901. [CrossRef]

38. Müller, J.; Böscke, T.S.; Schröder, U.; Müller, S.; Brauhaus, D.; Böttger, U.; Frey, L.; Mikolajick, T. Ferroelectricity in Simple Binary ZrO₂ and HfO₂. *Nano Lett.* **2012**, *12*, 4318–4323. [CrossRef] [PubMed]

39. Shimizu, T.; Katayama, K.; Kiguchi, T.; Akama, A.; Konno, J.; Sakata, O.; Funakubo, H. The demonstration of significant ferroelectricity in epitaxial Y-doped HfO₂ film. *Sci. Rep.* **2016**, *6*, 32931. [CrossRef] [PubMed]

40. Shimizu, T.; Katayama, K.; Kiguchi, T.; Akama, A.; Konno, T.J.; Hiroshi Funakubo, H. Growth of epitaxial orthorhombic YO1.5-substituted HfO2 thin film. *Appl. Phys. Lett.* **2015**, *107*, 032910. [CrossRef]

41. Phillips, J.C. *Bonds and Bands in Semiconductors*; Academic Press: New York, NY, USA, 1973.

42. Hagino, S.; Yoshio, K.; Yamazaki, T.; Satoh, H.; Matsuki, K.; Onodera, A. Electronic ferroelectricity in ZnO. *Ferroelectrics* **2001**, *264*, 235. [CrossRef]

43. Grants, F.A. Properties of Rutile (Titanium Dioxide). *Rev. Mod. Phys.* **1959**, *31*, 646–674. [CrossRef]

44. Abrahams, S.C.; Bernstein, J.L. Rutile: Normal Probability Plot Analysis and Accurate Measurement of Crystal Structure. *J. Chem. Phys.* **1971**, *55*, 3206–3211. [CrossRef]

45. Parker, R.A. Static Dielectric Constant of Rutile (TiO2), 1.6–1060°K. *Phys. Rev* **1961**, *124*, 1719. [CrossRef]

46. Von Hippel, A.R. *Dielectric Materials and Applications*; John Wiley & Sons: New York, NY, USA, 1954.

47. Gronschorek, W. X-ray charge density study of rutile (TiO₂). *Z. Kristallogr.* **1982**, *160*, 187–203. [CrossRef]

48. Sakata, M.; Uno, T.; Takata, M.; Mori, R. Electron Density in Rutile (TiO₂) by the Maximum Entropy Method. *Acta Crystallogr.* **1992**, *B48*, 591–598. [CrossRef]

49. Montanari, B.; Harrison, N.M. Pressure-induced instabilities in bulk TiO_2 rutile. *J. Phys. Condens. Matter* **2004**, *16*, 273. [CrossRef]

50. Bousquet, E.; Spaldin, N.A.; Ghosez, P. Strain-Induced Ferroelectricity in Simple Rocksalt Binary Oxides. *Phys. Rev. Lett.* **2010**, *104*, 037601. [CrossRef] [PubMed]

51. Glinchuk, M.D.; Khist, V.; Eliseev, E.A.; Morozovska, A.N. Ferroic properties of nanosized SnO_2. *Phase Transit.* **2013**, *86*, 903–909. [CrossRef]

© 2017 by the authors. Licensee MDPI, Basel, Switzerland. This article is an open access article distributed under the terms and conditions of the Creative Commons Attribution (CC BY) license (http://creativecommons.org/licenses/by/4.0/).

MDPI AG

St. Alban-Anlage 66

4052 Basel, Switzerland

Tel. +41 61 683 77 34

Fax +41 61 302 89 18

http://www.mdpi.com

Crystals Editorial Office

E-mail: crystals@mdpi.com

http://www.mdpi.com/journal/crystals

www.ingramcontent.com/pod-product-compliance
Lightning Source LLC
Chambersburg PA
CBHW051910210326
41597CB00033B/6094